11·50   UL/68813

KU-208-311

ENVIRONMENT AND MAN
VOLUME NINE

The Biological
Environment

Titles in this Series

ENVIRONMENT AND MAN
VOLUME NINE

The Biological Environment

General Editors

John Lenihan
O.B.E., M.Sc., Ph.D., C.Eng., F.I.E.E., F.Inst.P., F.R.S.E.

Director of the Department of Clinical Physics and Bio-Engineering, West of Scotland Health Boards, Professor of Clinical Physics, University of Glasgow.

and

William W Fletcher
B.Sc., Ph.D., F.L.S., F.I.Biol., F.R.S.E.

Professor of Biology and Past Dean of the School of Biological Sciences, University of Strathclyde; President of the Scottish Association for Biological Education.

UNIVERSITY LIBRARY
2 0 MAY 1981
LANCASTER

BLACKIE
Glasgow and London

Blackie & Son Limited
Bishopbriggs
Glasgow G64 2NZ

Furnival House
14-18 High Holborn
London WC1V 6BX

© 1979 Blackie & Son Ltd.
First published 1979

All rights reserved.
No part of this publication may be reproduced,
stored in a retrieval system, or transmitted,
in any form or by any means,
electronic, mechanical, recording or otherwise,
without prior permission of the Publishers

International Standard Book Numbers

Paperback 0 216 90749 7

Hardback 0 216 90750 0

Printed in Great Britain by
Thomson Litho Ltd., East Kilbride, Scotland

80 010768

Background to Authors

Environment and Man: Volume Nine

WILLIAM W. FLETCHER, B.Sc., Ph.D., is Professor of Biology at the University of Strathclyde, President of the Scottish Association for Biological Education, and Executive Editor and Chairman of the Editorial Board of the Royal Society of Edinburgh. From 1968 to 1971 he was Dean of the School of Biological Sciences, University of Strathclyde.

G. AINSWORTH HARRISON, M.A., B.Sc., D.Phil., is Professor of Biological Anthropology at the University of Oxford. He is a former President of the Royal Anthropological Institute and has undertaken field work in South America, Africa, New Guinea and Oxfordshire.

FORBES W. ROBERTSON, B.Sc., Ph.D., D.Sc., F.I. Biol., F.R.S.E., is Professor of Genetics at the University of Aberdeen, and European Editor of *Biochemical Genetics*. Prior to 1970 he was a member of the A.R.C. Unit of Animal Genetics at the University of Edinburgh.

JOHN LENIHAN, O.B.E., M.Sc., Ph.D., C.Eng., F.I.E.E., F.Inst.P., F.R.S.E., is Director of the Department of Clinical Physics and Bio-Engineering, West of Scotland Health Boards, and Professor of Clinical Physics, University of Glasgow.

J. E. SMITH, D.Sc., F.I. Biol., F.R.S.E., is a Personal Professor of Applied Microbiology at the University of Strathclyde, and Vice Dean of the School of Biological Sciences there.

JOHN N. R. JEFFERS, F.I.S., F.I. Biol., M.B.I.M., is Director of the Institute of Terrestrial Ecology, one of the component Institutes of the Natural Environment Research Council, U.K. He is a consultant to the UNESCO *Man and the Biosphere* programme and to the Scientific Committee on Problems of the Environment.

Series Foreword

MAN IS A DISCOVERING ANIMAL—SCIENCE IN THE SEVENTEENTH CENTURY, scenery in the nineteenth and now the environment. In the heyday of Victorian technology—indeed until quite recently—the environment was seen as a boundless cornucopia, to be enjoyed, plundered and re-arranged for profit.

Today many thoughtful people see the environment as a limited resource, with conservation as the influence restraining consumption. Some go further, foretelling large-scale starvation and pollution unless we turn back the clock and adopt a simpler way of life.

Extreme views—whether exuberant or gloomy—are more easily propagated, but the middle way, based on reason rather than emotion, is a better guide for future action. This series of books presents an authoritative explanation and discussion of a wide range of problems related to the environment, at a level suitable for practitioners and students in science, engineering, medicine, administration and planning. For the increasing numbers of teachers and students involved in degree and diploma courses in environmental science the series should be particularly useful, and for members of the general public willing to make a modest intellectual effort, it will be found to present a thoroughly readable account of the problems underlying the interactions between man and his environment.

Preface

OUR BIOLOGICAL ENVIRONMENT PROBABLY HAD ITS ORIGINS A FEW thousand million years ago when radioactivity from the rocks acting on the gases of the primitive atmosphere resulted in the production of self-replicating molecules. From such simple beginnings we now have 4000 million people and a vast array of plant and animal life.

Professor W. W. Fletcher takes us step by step along the evolutionary trail, describing processes (sometimes factual, sometimes conjectural) by which we have reached our present-day flora and fauna. A biological lesson to be learned from the mammals and seed plants is that evolutionary success and survival depends largely on protection of the embryo.

Professor Ainsworth Harrison continues with the evolution of man, from his arboreal ancestry in tropical Africa to modern *Homo sapiens*. He notes how the environment has acted in two distinct but inter-related ways: first by moulding the phenotype in individual development through interaction with the individual's genetic constitution, and secondly in favouring certain phenotypes in survival and reproduction. He emphasizes the importance to the early hominids of bipedalism, leaving the forelimbs free for carrying and eventually using tools. With scholarship and clarity, Professor Harrison describes the characteristics and mode of life of *Australopithecus*, *Ramapithecus*, *Homo habilis* and *Homo erectus*. Most successful of all was (and is) *Homo sapiens*.

In his chapter on natural selection, Professor Forbes Robertson looks first at the evidence for physical and physiological adaptation in contemporary man. (Why, for example, are northern people predominantly white? How does the South African Bantu tolerate high temperature?) He goes on to examine genetic differences that occur among humans, covering topics such as sickle-cell anaemia and blood groups. His discussion of selection and the future is particularly thought-provoking.

Professor John Lenihan examines the concept of man as a machine. Until recently the mechanical improvement of man was limited to external aids such as spectacles and artificial limbs. Now, however, we have plastic heart valves, artificial kidneys and heart-lung machines. The imitation of the biological environment is, in the long term, one of the greatest challenges facing man's ingenuity and understanding. Professor

Lenihan reminds us that progress is not inevitable, and suggests that the attack on this particular problem may not be successful until engineers forget their justifiable pride in the achievements of technology and adopt the humbler attitudes of the biologist.

Since the time of Louis Pasteur we have become increasingly aware of the importance of microbes. Early emphasis was put on their role as agents of disease in plants, animals and man; later we came to understand their importance in Nature's economy; and recently we have begun to appreciate their industrial value. Professor John Smith deals comprehensively with these themes, pointing out interesting aspects of the social, biological, and industrial interactions between man and the microbes that share the earth with him.

Finally Mr J. N. R. Jeffers discusses the reasons for, and methods of, biological conservation, which has emerged as one of today's principal political, social and economic issues. He examines the ecological basis of our understanding of the biological environment, considers some of the strategies which have been used for conservation nationally and internationally, and discusses in detail the genetic basis for conservation. Man's perception of the basic problems greatly affects the extent of the impact that he will have upon the environment, and will ultimately depend on the extent to which he is successful in predicting the results of his manipulations of marine, freshwater and terrestrial systems.

This is a book of special pertinence to students and lecturers in human biology and ecology, evolutionary biology, physiology, genetics and conservation.

Contents

CONTENTS

CHAPTER ONE

EVOLUTION OF THE NATURAL ENVIRONMENT

W. W. FLETCHER

In the beginning

Astronomers tell us that there may be more than 100 billion* galaxies in our Universe and that each galaxy may contain more than 100 billion stars. The Milky Way, a lens shape some 100 000 light years† across, is but one of these galaxies and our Sun is but one of its stars which, with the attendant planets Mercury, Venus, Earth, Mars, Jupiter, Saturn, Uranus, Neptune and Pluto, make up our Solar System.

So far life has been found only on Earth, and it is unlikely to be found on the other planets of the Solar System. On a statistical basis, however, it does seem possible that the Universe could be teeming with life, and it has been estimated that within the Milky Way alone there may be between 100 million and 100 billion cold planets, associated with stars, which could support living things. "Life", says the eminent American biologist George Wald, "is a cosmic event".

How could such life have come into being?

It is not our present purpose to discuss the various theories that have been proposed to account for the origin of the galaxies and their component stars. What is relevant, however, is that it is now generally accepted that the Earth was formed by a process of "accretion", whereby a mixture of gas and dust from exploding supernovae‡ was picked up in a whirling motion until a considerable aggregate was formed. As these

---

* "Billion" is used in this chapter with the American meaning of one thousand million = 10^9.

† One light year is the *distance* that light, travelling at 186 000 miles per second, travels in one year.

‡ Supernova—an exploding star, e.g. the Crab Nebula which exploded in 1054 and whose remains are still visible.

aggregates, or "embryos" as they have been called, rotated around the Sun, they began to collide with each other to form larger aggregates, and thus planet Earth (and other planets) grew. The planets are still being bombarded by particles of interstellar dust to the extent of millions of tons per year, but in the case of Earth we are now protected by our atmosphere. A few meteorites do break through our protective barrier from time to time, and we pick up about fourteen tons of cosmic dust per year.

The original Earth was not a molten mass; it was a cold body which later became hot due to the activity of radioactive chemicals trapped inside it. The disintegration of these radioactive chemicals (each of which is transforming at a set unchanging pace completely unaffected by environment) enables us to date the Earth as around 5 billion years old.

As the Earth heated up, water, formed by the combination of free oxygen and hydrogen, boiled off from its surface, forming great masses of clouds which were probably hundreds of miles thick. As cooling set in, the rains began—torrential rains that may have lasted for centuries—eventually to form the oceans and the great river basins of the world. The atmosphere was not as we know it now. Free oxygen was absent because any available had been taken up by the metals, as they converted themselves to oxides, and by hydrogen to form water. The atmosphere of Earth was probably composed, as the atmospheres of the other planets of the Solar System are today, of ammonia, methane, water vapour, carbon dioxide and nitrogen, with traces of other gases.

The origin of life

It was within this mixture that living things came into being. We do not know how they were first formed, but some interesting theories have been proposed. One of the most philosophically pleasing theories, and one which seems to be gaining increased credence among scientists, is that put forward by the Russian A. I. Oparin, supported by the Briton J. B. S. Haldane, and shown to be scientifically possible by the Americans S. L. Miller and H. C. Urey.

In 1922 Oparin proposed that, under the influence of radiations, simple substances such as carbon dioxide, methane, ammonia, and water gave rise to more complex chemicals such as sugars, proteins and fats, and that these then combined in special ways to give rise to the first recognizable living thing. Oparin saw the progression as being

inorganic ⟶ organic ⟶ simple living things ⟶ complex living things

He viewed the evolution of these groups, not as separate phenomena, but as steps in the unfolding of one and the same process. He and Haldane believed that the gases of the primitive atmosphere, subjected to ionizing radiations, would split into their constituent parts and then recombine at random to yield a vast array of organic chemicals, including proteins, which would be washed down by the rains into shallow pools which would resemble a "hot dilute soup". These organic chemicals would be combined within "skins" of fat/protein complexes to form the first recognizable unicellular organisms.

These ideas remained theory until 1953 when Miller and Urey put them to experimental test. We know that the essential ingredients of living things are proteins (for structure, and the formation of enzymes), nucleic acids (for the storage and transmission of hereditary information), adenosine triphosphate (ATP) (for the storage and release of energy), water, carbohydrates, fats and miscellaneous other chemicals. Urey and Miller put a mixture of methane, ammonia and hydrogen together with water into a continuously circulating system and subjected the mixture to electrical discharges. They allowed the experiment to run for a week and then the resulting fluid was analysed. It contained an amazing number and variety of organic chemicals including amino acids (the building blocks of proteins). Other scientists who have followed similar procedures have produced in addition, purines and pyrimidines (the building blocks of nucleic acids) and carbohydrates. These experiments dramatically demonstrated that a great variety of organic "building blocks" of living things could, as Oparin and Haldane had predicted, be formed from the simple inorganic gases of the primitive atmosphere. Later experiments, e.g. by Fox in the United States and by Schramm in Germany, have demonstrated that using these organic materials it is possible to synthesize primitive proteins, ATP, and nucleic acids, by providing the conditions that probably prevailed in the ancient atmosphere. Furthermore, it was shown that these products could aggregate and be enclosed in thin membranes of protein and fat which resemble present-day cell membranes. Those chemicals that combined and enclosed themselves in this way, so the theory runs, constituted the first living organisms.

It has not, however, so far been possible to synthesize a piece of nucleic acid, say DNA (deoxyribonucleic acid), that has the power of self-reproduction. This must just be a matter of time, for already scientists have come remarkably close to doing so; for example, it is possible to "seed" a mixture of purines and pyrimidines with a small piece of DNA which then proceeds to build a chain of DNA from such raw materials. It is also possible to induce a piece of DNA to form primitive proteins when it is seeded into a mixture of amino acids, and it is possible to synthesize a chain of DNA *de novo* using known purines and pyrimidines.

The final step cannot be far away. When it does come, we will have demonstrated that it is possible for a living organism to come into being under the conditions that existed when the world was very young. This will not mean that this is how it did happen, though the presumptive evidence will be very strong.

In order to survive, these early organisms had to have food which they probably got in the form of other organic molecules which had not aggregated together to form cells. Such foodstuffs would be in relatively short supply and were possibly supplemented by "chemosynthesizers", i.e. organisms which could (as some bacteria still do today) derive energy from sulphur, iron, nitrogen and certain other metallic and non-metallic materials. This was, however, only a partial solution to the food and energy problem. There was a requirement for a steady bountiful supply of food and energy, and this could only come from the Sun. At some stage some of these organisms developed the green colouring matter chlorophyll, and by means of this chlorophyll, using carbon dioxide and water, and trapping the energy of sunlight (photosynthesis), they (we may now term them "plants") provided an inexhaustible supply of energy-rich organic molecules such as carbohydrates, fats and proteins, not only for themselves but also for the other organisms ("animals") which did not develop chlorophyll. From the fossil record it is clear that the primitive plants (algae) developed more than 3 billion years ago.

It should be noted that as yet there was no free oxygen in the atmosphere, as life could probably not have come into being other than under "reducing" conditions. The food molecules were therefore broken down by organisms by the process of fermentation (in much the same way as yeast breaks down sugars today) yielding carbon dioxide, alcohol and energy. This was not a particularly efficient system, since not all of the energy is released from the food (much still being bound up in the alcohol) according to the following simplified equation:

$$\underset{\substack{\text{glucose}\\(180\text{ g})}}{C_6H_{12}O_6} = \underset{\text{ethyl alcohol}}{2C_2H_5OH} + \underset{\text{carbon dioxide}}{2CO_2} + 50\,000\text{ cal}$$

What was now needed was oxygen, so that aerobic respiration could be substituted for fermentation—and providentially the process of photosynthesis not only supplied organic molecules, it also supplied oxygen (as seen in the following very simplified equation):

$$\underset{\substack{\text{carbon}\\\text{dioxide}}}{6CO_2} + \underset{\text{water}}{6H_2O} \quad \underset{\text{chlorophyll}}{\overset{\text{light}}{=}} \quad \underset{\text{glucose}}{C_6H_{12}O_6} + \underset{\text{oxygen}}{6O_2}$$

Using this oxygen, organisms (plants and animals) could dispense with the fermentation process and could break down foodstuffs much more efficiently to yield more energy (as most plants and animals—including man—do today).

$$C_6H_{12}O_6 + 6O_2 = 6CO_2 + 6H_2O + 700\,000 \text{ cal}$$

The only by-products of this process are carbon dioxide and water. These are not poisonous (unlike ethyl alcohol) and are easily disposed of in any environment. They can also be used over and over again in the process of photosynthesis.

Without photosynthesis life might have continued at best within a very narrow cycle. Even with photosynthesis, organisms could lead only a marginal existence with fermentation. The release of oxygen made respiration possible, which with photosynthesis made all things possible. George Wald has said:

It raised organisms above the subsistence level; it provided them with capital which they invested in the great enterprise of organic evolution.

It set the great evolutionary process in train which has resulted in the myriads of organisms that occupy our oceans, fresh water, air and land of today. It was to culminate in man.

The entry of oxygen into the atmosphere also liberated organisms from the bondage of living under water. The Sun's radiations are composed partly of ultra-violet rays which are lethal to living organisms. If such radiations were to reach the Earth in strength, then life would cease except under water, which acts as a filtering agent. Some of the oxygen rose high in the atmosphere and was converted to ozone by these self-same radiations and formed a protective layer, ozone, which filtered out most of the harmful radiations. The way was now open for organisms to come out of the water and live on the land. That step was still a long way ahead, but the conditions had come into being to make it possible.

As photosynthesis increased, it brought about fundamental changes in the atmosphere. As we have noted, before photosynthesis there was no free oxygen. Now increasing amounts of free oxygen were being released into the ocean and from there into the atmosphere. There some of it reacted with methane, transforming it into carbon dioxide and water:

$$\underset{\text{methane}}{CH_4} + \underset{\text{oxygen}}{2O_2} \longrightarrow \underset{\substack{\text{carbon}\\\text{dioxide}}}{CO_2} + \underset{\text{water}}{2H_2O}$$

It also reacted with ammonia and with any cyanide present to form molecular nitrogen, water and carbon dioxide:

$$4NH_3 + 3O_2 \longrightarrow 2N_2 + 6H_2O$$
$$4HCN + 5O_2 \longrightarrow 2N_2 + 2H_2O + 4CO_2$$

The atmosphere was thus cleared of methane, ammonia and any cyanides present, and replaced by nitrogen, water vapour, carbon dioxide and, in time, free oxygen. Thus the ancient atmosphere was gradually transformed into the modern one.

Reproduction

The success of the cellular mode of construction may be assessed by the fact that within a few million years of the first vital tentative steps having been taken, the oceans were teeming with living things. The vast numbers were due to the reproductive capacity of the nucleic acids; the variety was due to the adoption of the sexual mode of reproduction, coupled with the ability of the nucleic acids to make mistakes in their duplicating processes (mutation). DNA, the material carrying the hereditary message, may inexplicably alter its structure in the duplicating process. Such structural changes may alter the characteristics of the resulting organism. The change is termed a *mutation* and the resulting organism a *mutant*. Such mutants are the stuff of evolutionary change.

The method of reproduction of the early organisms was a simple one—they split into halves, which then separated, and thus there were two organisms where formerly there had been one. Organisms produced by this method were genetically alike—apart from the odd mutation.

The origins of sexual reproduction are obscure, but it is certainly true that the fusion of two unicellular organisms (or parts of multicellular organisms) was the spark that set off the whole evolutionary process. Instead of reproducing by simply dividing, some organisms either fused together or produced special cells (sex cells) which joined together; the resulting zygotes then divided again and again to form new individuals. Sexual reproduction was a striking success, and it was the pathway that nearly all organisms followed. Those organisms that did not develop it remained low in the evolutionary scale. It was successful because the joining of two cells, each with its own hereditary make-up, led to a completely different individual being produced. Sexual reproduction increased the variability within a species, and natural selection went to work to secure the advancement of the species.

A concurrent advance was the development of the multicellular body. Unicellular organisms can be highly efficient—the form has persisted in a number of species down to the present day—but the future lay in large measure with the multicellular form. Perhaps some unicellular forms failed to separate when they divided, so that a cluster of cells was formed; perhaps some unicellular forms joined together to form a multicellular mass. Whatever the mechanism, the form had evident evolutionary advantage. With the emergence of the multicellular organism and its consequent complexity, certain cells (and the tissues formed from them) took on specific tasks. They ceased to be generalized cells. Some took on the function of reproduction, others excretion, still others message transmission and so on. Thus, in time, complex living things evolved with a sharing out of various tasks within the body.

Plant life in the oceans

In time, the oceans teemed with life as organisms found it a superb medium in which to proliferate. Beginning, no doubt, in the warmer pools, they gradually extended their range until they were to be found world-wide. Among the simplest were and are the microscopic diatoms, of which today there are more than 5000 species. Diatoms contain large amounts of silicon and as a result of this they leave splendid fossil remains. There seems to have been little change in their structure over millions of years, indicating perhaps the uniformity of their environment. Among the plants there evolved the blue-green algae (Cyanophyta), the green algae (Chlorophyta), the red algae (Rhodophyta) and the brown algae (Phaeophyta)—all relatively simple plants, although in size they were in time to range from the microscopic of the blue-green algae to massive brown algae such as *Nereocystis* and *Macrocystis* which today attain lengths of 50–70 metres and which form great underwater "forests". Free-floating forms such as *Sargassum* formed dense "forest" beds some 20 metres in depth, thus providing safe breeding and grazing areas for the animals of the sea. These algae, as well as developing diverse forms, also evolved various pigments which enabled them to colonize depths that might otherwise have been inaccessible due to low light intensities. Thus, in addition to chlorophylls, the blue-greens contain the accessory pigment phycocyanin; the browns have the accessory xanthophyll pigment, fucoxanthin; and the reds the accessory phycobilin pigments such as r-phycoerythrin (red) and r-phycocyanin (blue). The blue-greens, of which some 1500 species are known, have been successful in colonizing the land, being found on rocks and in soils. They are remarkably versatile. Many are able to resist desiccation, high temperatures (e.g. in hot springs up to 85°C), low temperatures (found growing on snow) and even desert areas. The latter colonization is no doubt aided by the ability of some of the blue-greens to "fix" and utilize atmospheric nitrogen.

Animal life in the oceans

In the sea, animal life was evolving too. Among the simplest and most primitive forms were the sponges, corals, sea-anemones and jellyfish. In

Table 1.1   Geological eras, periods and their plants and animals

Era	Period		Duration (millions of years)	Time from present (millions of years)	Some plant and animal developments
CAINOZOIC (Age of mammals)	Quaternary	Recent Pleistocene	1+	1+	Races of man First man
	Tertiary	Pliocene	69	70	Ape-like and man-like creatures
		Miocene			Monkeys and ancestors of apes and men
		Oligocene Eocene/ Palaeocene			Spread of birds Modern mammals and flowering plants
MESOZOIC (Age of Reptiles)	Cretaceous		65	135	Spread of flowering plants Extinction of dinosaurs Egg-laying and marsupial mammals
	Jurassic		45	180	First flowering plants Dinosaurs dominant, first primitive birds, insects abundant Continents forming from Pangea
	Triassic		45	225	Cone-bearing trees (gymnosperms) Flying and water reptiles, dinosaurs, mammal-like reptiles
PALAEOZOIC	Permian		45	270	Cone-bearing trees, reptiles abundant
	Carboniferous		80	350	Forests of tree ferns and lycopods and seed ferns; amphibians invade land; first reptiles
	Devonian		50	400	Ferns and lycopods; cartilaginous and first bony fishes, lung-fishes. Insects invade the land
	Silurian		40	440	First land plants; jawless fish. Arachnids invade the land
	Ordovician		60	500	Corals; first primitive vertebrates

Table 1.1 *contd.*

Era	Period	Duration (millions of years)	Time from present (millions of years)	Some plant and animal developments
PALAEOZOIC *contd.*	Cambrian	100	600	Sponges, corals, marine worms, seaweeds, sea anemones, jellyfish
ARCHAEOZOIC PROTEROZOIC	Pre-Cambrian	(approx.) 4400	(approx.) 5000	Algae and some marine worms. Fossils rare, but some algae have been found which are approximately 2600 million years old

the latter, the nervous system had begun to develop by means of two inter-connecting nerve nets—one controlling swimming movements, the second feeding and other movements. There are few fossil records of jellyfish because of the soft texture of their bodies, but some impressions have been found in pre-Cambrian rocks dating back more than 600 million years. Moving up the evolutionary scale we come to the marine worms (the most highly developed forms of which were segmented, had blood vessels, a highly developed nervous system, gills and eyes), crabs, lobsters, snails (whose ancestors go back to the pre-Carboniferous period), clams and mussels (we find examples of these animals in Ordovician rocks of 500 million years ago), squids and octopuses. Among the group known as the echinoderms are the brittle stars, starfish, sea urchins, sea lilies and sea squirts (tunicates). All of the foregoing animals are invertebrates, i.e. they do not possess a backbone.

For the origin of the vertebrates, according to some scientists, we have to look closely at the sea squirts (figure 1.1). These little animals consist of little more than a covering or tunic, inside which there is a filtering chamber with a small stomach, intestine and reproductive organs. Most of them stay attached to the sea bed, filtering their food out of the water. Improbable as it seems, these little creatures (according to one theory) were the forerunners of the vertebrates that were to lead eventually to man. Sitting on the sea floor filtering water with little expenditure of energy is a fairly efficient way of food gathering but, when it came to extending its territory, the fixed form was at a serious disadvantage. Some of them, however, developed a motile stage in their life cycle. This little "tadpole" was able to swim away, find new

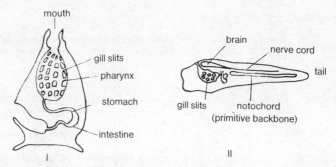

Figure 1.1 Two stages in the life history of a sea squirt. I. immobile filtering stage
II. motile larval stage which may have been the forerunner of the vertebrates.

unoccupied territory, settle down and develop into a fixed sea squirt. Some
sea squirts still do this today. Here, however, is a strange twist. Some of these
"tadpoles", instead of turning into sea squirts, continued an independent
existence as tadpoles and developed sexual organs. The first little primitive
"fish" had come into existence, evolving from the motile stage of the sea
squirt. They were simple in structure—gill slits for breathing, a nerve cord
running along the back of the body, and a long tubular structure (the
notochord) was the forerunner of the backbone (notochords can be seen in
the embryos of all the higher vertebrates, indicating the ancestry of the spinal
column). The theory may or may not be correct. It is certainly an interesting
one.

Plants move to the land

Up to the time of the Silurian Period (some 440 million years ago, see
Table 1.1), the land was probably barren except for the presence of a few
blue-green algae. If, as seems likely, the first living things came into being
in the oceans some 3000 million years ago, then for almost all of this period
the land remained fairly sterile. During this period plants and animals were
evolving in the oceans, but they did not invade the land. Why should they?
The obstacles were formidable. Life in the oceans was fairly comfortable,
and it was not until numbers became unbearable that the evolutionary
pressure went to work. Living on land was a very different prospect from
living in water. We do not know what these early pioneer plants looked
like, but they must have been very adaptable. In the sea they were surrounded by water containing ample mineral salts; they had no need of
transport, and so specialist organs and tissues were either not evolved at
all or, if evolved, were primitive and simple. On the land, on the other
hand, plants faced the possibility of desiccation, and only those with the
potential for evolving water-proofing tissues, absorbing structures, con-

ducting tissues, ventilating cells and supporting structures were able to exploit the new environment. It seems likely that the green algae were the ancestors of the land plants—their chlorophylls correspond closely, and the greens are the only algae to store reserve food material as starch (as the land plants do today).

The first animal invasion of the land—the arachnids

The first invasion of the land by animals also probably took place more than 450 million years ago when scorpion-like creatures made their way out of the water. Certainly we have fossils of these creatures dating from Silurian times. The danger of drying out was met by the possession of a hard exoskeleton which also provided body support.

One major problem was the changeover from extracting oxygen from water to absorbing it from the air, but here again this group was well placed to bring about the change in that it required relatively minor mutations to bring the external leaf-like gills, which were situated beneath the abdomen, into the body cavity. From these primitive scorpion-like forms, other arachnids arose, giving rise to modern-day spiders, mites and ticks.

The second animal invasion of the land—the insects

The insects differ from the arachnids in that they have three pairs of walking legs (the arachnids have four pairs) and the body is segmented into a head, thorax and abdomen. Almost all forms have wings mounted on the thorax and have compound eyes made up of thousands of little lenses. The origin of the insects is not clear, but they probably arose from marine-living centipedes. Unlike the arachnids they had no gills and they did not depend on blood for the transportation of oxygen and carbon dioxide. Instead, the insects developed little pores on the sides of the body from which a series of little tubes, termed *tracheae*, branch out carrying gases to and from all parts of the body.

This movement of insects on to the land was to have fundamental consequences for man's environment. Without it we would have no flowering plants, fewer diseases, and more food. Here we may note that primitive insects have been found as fossils in Devonian rocks (400 million years ago).

Continuing evolution in the sea

There are living today a number of small sea creatures which are not far removed from the primitive tadpole—the forerunner of the vertebrates

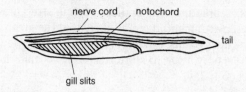

Figure 1.2 *Amphioxus*—a creature living today which might form a link between the squirt larva and the vertebrates.

that we have been discussing. One such creature is *Amphioxus* (figure 1.2). It looks like a small minnow, but it has no backbone and no brain, only a nerve cord running along its back. It has a well-developed notochord and about 50 pairs of gill slits. It is capable of swimming, but spends most of its life stuck down in the sand by its tail-end, filtering water through its many gills, thus revealing the possibility of its filter-feeding ancestry. The group to which *Amphioxus* belongs is a step towards the true fish. We can trace yet another step by looking at the lampreys. These are about two feet in length. They have no jaws but only an adhesive oral sucker by which they attach themselves to other fish. Possibly they are degenerate forms, but equally possibly they indicate a line that was to lead to true fish, because in the fossil record we find many jawless fish (figure 1.3) which fed

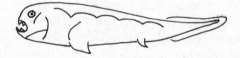

Figure 1.3   Armour-plated jawless fish of the Devonian Period.

by sucking up debris from the sea bed. These fossil fish had no fins, and the whole body was encased in armour made up of bony plates. They moved by means of undulations of the body. If we judge by the number of fossils found, they were abundant some 400–500 million years ago but, since they were armoured-plated, they did leave very good fossils.

Fish evolution

As we move from the Silurian into the Devonian Period (Table 1.1), we find that the armour plating of the jawless fish was being replaced by fine scales which gave the fish greater mobility. They were becoming dependent on speed for their survival, and paired fins assisted the process. The development of jaws made them more efficient in catching their prey, and the fins gave them control of their movements in the water. In the primitive fish the skeleton was composed of soft rubbery cartilage (some representatives, such as sharks, are still found with such skeletons today) but in the evolutionary process this cartilage was being replaced by tough bone. This line was to give rise to the wide variety of bony fish that inhabit the seas today.

The first tentative steps were also being taken that would enable animals to live out of water. Water is 800 times denser than air, and the oxygen in it is 30 times more dilute than in air. The fast movements were now demanding more available energy and, in order to meet this demand, many fish were augmenting their supply of dissolved oxygen by taking in mouthfuls of air from the atmosphere. The Indian perch of today has special air chambers which enable it to come out of the water for quite considerable spells. It comes to the surface to gulp air, and may actually drown if prevented from doing so. The ability to gulp air was no doubt an important factor in the survival of certain fish in the face of drought. By chance mutation some developed a pair of pouched outgrowths from the pharynx. These were the forerunners of lungs, and they were inflated on the intake of air. By reading the fossil record we can deduce that some of these "lung fishes" burrowed into the mud by means of lobed fins during periods of drought. They were almost ready to give up their total dependence on an aquatic environment and to move on to the land.

The coelacanth

In 1938 a strange fish was brought up out of the depths off the east coast of South Africa (figure 1.4). It was recognized as a coelacanth—one of a member of a group that was thought to have been extinct for some 70 million years. Despite organized searches it was not until 1952 that another specimen was pulled out of the water near the Mozambique Channel. Since that time, more have been recovered, and a close study of them indicates that these fish may be "living fossils"—part of the group that in the distant past crept on to the land to become the first amphibians. They have both gills and rather primitive lungs; the fins are borne on a scaly stalk protruding from the body, and each fin articulates through a

Figure 1.4 Coelacanth, a living fossil. Note that fins are borne on short stalks which could be the forerunners of limbs.

single structure, just as the limbs of the frog, bird, dog or man hinge on to the humerus of the arm or the femur of the leg. The fin has become a little limb. Relatives of the coelacanth developing such structures some 350 million years ago were ready to move on to the land to become the first amphibians.

The third animal invasion of the land—the amphibians and their environment

The amphibians (figure 1.5) could not leave the water completely. They lived part of their lives on the land, but they had to return to the water to lay their eggs, and they could not stand long exposure to dry air. In

Figure 1.5   An early amphibian of the Devonian Period.

addition to the development of lungs and limbs, the amphibians developed two more structures that were vital to their survival. The first was a third chamber in the heart which increased its efficiency as a pumping organ. The atrium developed a partition; the right atrium received deoxygenated blood from the body, while the left atrium received oxygenated blood from the lungs. Both fed into a single ventrical (figure 1.6).

The second development was the modification of one pair of gills as openings which became covered with a membrane to form ears. Hearing was, and is, of much greater importance on land than in water. In one respect,

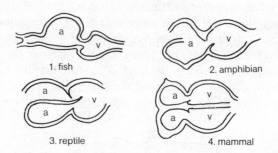

Figure 1.6   Stages in the development of the heart. Note the increasing separation of the atrium (a) into two chambers and ventricle (v) into two chambers as we progress from (1) fish to (2) amphibian to (3) reptile to (4) mammal. This reduces the mixing of oxygenated and deoxygenated blood (not to scale).
(after *Biology* by Cockrum, McCauley & Youngren, W. B. Saunders Company, Philadelphia)

however, the amphibians did not show much change from their fish ancestors—the brain and nervous systems remained more or less as they were.

What sight greeted these first vertebrates who came on land? This was the beginning of the Carboniferous Period (it lasted more than 80 million years, see Table 1.1) when the vegetation was made up of large lycopods (which had evolved from early Devonian times). They reached tree stature, with the trunks more than one metre in diameter and exceeding 35 metres in height. Many had crowns of branches at the top bearing slender cylindrical spore-bearing organs. Within the swamps where they lived they were to form the major part of our present-day coal measures. Most of them died out, due to the cool dry conditions that set in towards the end of the Carboniferous Period, leaving only small delicate herb-like *Lycopodium* and *Selaginella* to come down as representatives to the present day. Articulates (the only present-day form is *Equisetum*—horse tails) also reached their peak in this Period. Many of them reached up to 10 metres in height and were very conspicuous with their leaves and branches borne in whorls. Mosses and liverworts were also abundant.

All of these plants reproduced by spores, much as the ferns do today. The seed habit (see page 16) had not yet been developed.

This was essentially a swamp flora as these plants had not yet evolved sufficiently for their reproduction processes to become independent of external water. The temperatures were semi-tropical and the plant growth was luxuriant.

Fossils indicate that some of the amphibians were up to eight feet in height, and we can picture them floundering around in the swamps, insecure in their tenure of the land.

The reptiles and their environment

Some time during this Carboniferous Period some of the amphibians were evolving into reptiles which were eventually to displace them and for a period rule the earth. To what do we attribute their success? As we have seen earlier, in the coelocanths, relatively small changes can have tremendous consequences. In the case of the reptiles, the development of a protective coating for the egg gave them independence from their watery environment. Eggs could now be laid on dry land without the danger of their drying out. Indeed, this development of an egg coat is the main difference between the amphibia and the reptiles. As the Carboniferous Period (some 350 million years ago) gave way to the Permian (270 million years ago, see Table 1.1), the reptiles were abundant and the amphibians, unable to compete, were in steady decline.

The development of an impervious egg-covering had another fundamental consequence—it led to sexual intercourse between individuals.
Since the reptilian egg was impervious to water, it was also impervious to
sperm, so that henceforth fertilization could only be effected by the
injection of sperm into the female before the shell formed (amphibian eggs
were fertilized after being shed from the female). Within the shell the
embryo was surrounded by a membrane—the amnion—which held a
small amount of water. The embryo is really an aquatic organism. This
situation remained practically unchanged in the birds which, as we shall
see, evolved from the reptiles.

As the amphibians disappeared, so too did the lycopods and the
articulates, which were replaced by the Pteridophyta (the ferns) and the
pteridosperms (the seed ferns). The former have many living representatives today; the latter are known only as fossils. Both reached their
greatest structural diversity during the late Carboniferous Period. The
ferns have evolved and persisted down to the present day and, although
they have never become dominant, in many places they form a conspicuous part of our environment. The bracken fern (*Pteridium aquilinum*)
colonizes large tracts of moorland in the United Kingdom; in the tropics
and in warm temperate regions "tree ferns" are prominent. At the present
day there are eight orders of the Pteridophyta, one of which (the Filicales)
contains by far the greatest number of existing ferns with twelve families,
about 170 genera and nearly 9000 species.

The pteridosperms, which are known only as fossils, reached their
maximum both numerically and structurally during the late Carboniferous Period. Until fairly recent times (beginning of twentieth century)
many of the leaves found in the Carboniferous rocks were believed to be
true ferns, and the Period was referred to as "The Age of Ferns"; but,
unlike the ferns, they bore ovules and seeds, and their line gave rise to the
gymnosperms and the angiosperms.

The seed habit

One of the major evolutionary steps forward in the Plant Kingdom was
the development of the *seed habit*. The ferns and their allies show well-
marked "alternation of generations" in that the life cycle is completed by
the production of two distinct life forms—the sporophyte and the gametophyte. The sporophyte (which is large with prominent leaves) is diploid
and, by a process of meiosis, it forms haploid spores which are borne
within little sacs known as *sporangia*. On ripening, these spores are shed
and germinate, each giving rise to the gametophyte generation (a minute
heart-shaped plant in the case of the fern) which develops male and

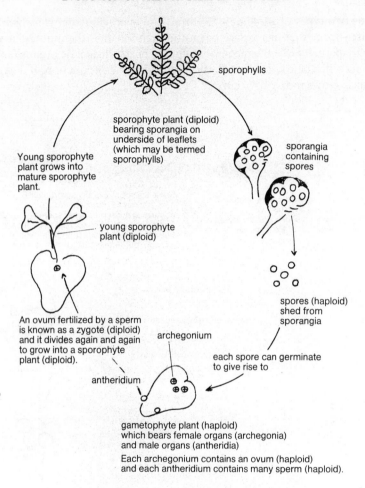

sporophylls

sporophyte plant (diploid)
bearing sporangia on
underside of leaflets
(which may be termed
sporophylls)

Young sporophyte
plant grows into
mature sporophyte
plant.

sporangia
containing
spores

young sporophyte
plant (diploid)

spores (haploid)
shed from
sporangia

An ovum fertilized by a sperm
is known as a zygote (diploid)
and it divides again and again
to grow into a sporophyte
plant (diploid).

archegonium

each spore can germinate
to give rise to

antheridium

gametophyte plant (haploid)
which bears female organs (archegonia)
and male organs (antheridia)
Each archegonium contains an ovum (haploid)
and each antheridium contains many sperm (haploid).

Figure 1.7   Life cycle of a fern.

female sexual organs—antheridia and archegonia respectively. Sperm
escape from the antheridia, swim towards the archegonia, and there
fertilize the egg cell which is contained within each archegonium. The
resulting fertilized egg, which is now diploid and known as a zygote,
divides repeatedly by the process of mitosis to give rise to a sporophyte
plant. Thus the cycle is completed (figure 1.7).

In some of the relatives of the fern, e.g. *Selaginella* (figure 1.8), there is
an evolutionary advance in that two types of spore, one small (microspore) and one large (megaspore), are produced by the sporophyte and

thus two types of gametophyte ensue. The microgametophyte which has arisen from the microspore bears the male organs (the antheridia with their sperm), while the megagametophyte bears the female organs (archegonia with their ova). Fertilization results in the production of a zygote which grows into a sporophyte.

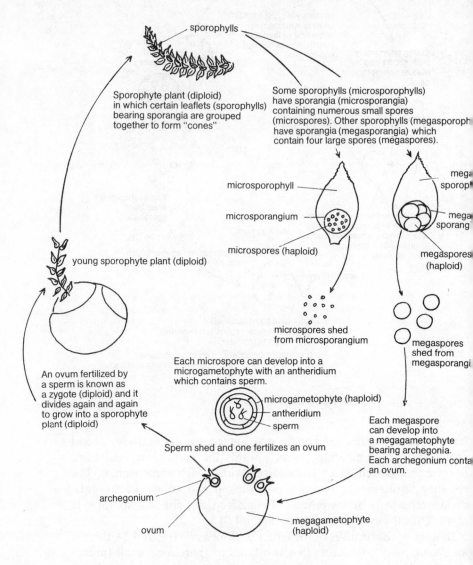

Figure 1.8   Life cycle of *Selaginella*.

There is the obvious danger in these life cycles of the spores being shed into hostile environments and thus the life cycles being broken. Better by far if the microspores and megaspores could be retained on the sporophyte and develop there. And this is what the "seed ferns" (and their descendants, the gymnosperms and the angiosperms) did. Each megasporangium enclosing a megaspore became enclosed in a covering or integument. The whole, now known as an *ovule*, was borne on a specialized leaf termed the megasporophyll or *carpel*. Each microsporangium (or pollen sac) enclosing the microspores (or pollen grains) was borne on a special leaf called the microsporophyll or *stamen*. The pollen grains, on being shed, were transported (by wind or insects) to the ovule, where they released a sperm (or male nucleus) which fused with the female ovum within the ovule. The resulting zygote, still within the ovule, developed into an embryo plant. The ovule ripened with this little embryo plant still within; the whole formed a seed. This mechanism proved to be very successful. It gave rise to the gymnosperms (figure 1.9) and, more importantly, to the flowering plants (figure 1.10).

The gymnosperms

In the gymnosperms (literally "naked seed") the ovules are situated on, but not enclosed by, the carpels (see figure 1.9). In some orders the carpels may be grouped together to form cones (the female cones). Similarly the stamens may be grouped together to form male cones.

As the Palaeozoic Era gave way to the Mesozoic, the great fern-like trees were on the decline and were being replaced by the seed-bearing gymnosperm trees. In the animal kingdom, the reptiles were also replacing the amphibians. Indeed, the Mesozoic Era (beginning some 250 million years ago and lasting for about 150 million years) has been called the Age of the Gymnosperms and the Age of the Reptiles. Great geological events were taking place as the Earth's crust contorted to form great ranges of mountains, many of which were subsequently to be eroded to form hills.

Two of the most primitive groups of the gymnosperms are the Bennettitales (which did not form cones) and the Cycadales (which did).

The Bennettitales, which are generally believed to have originated from the pteridosperms, are known now only as fossil remains. They reached their zenith in the Jurassic, and declined dramatically to extinction in the late Cretaceous Period. Indeed they parallel the rise and fall of the dinosaurs, suggesting that the same conditions suited both. It has even been suggested that they were interdependent, but there is no evidence for this. The Cycadales, which together with the Bennettitales formed the dominant part of the flora in Mesozoic times, have persisted down to the

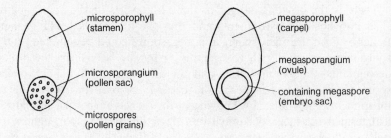

Figure 1.9   Gymnosperm reproduction.

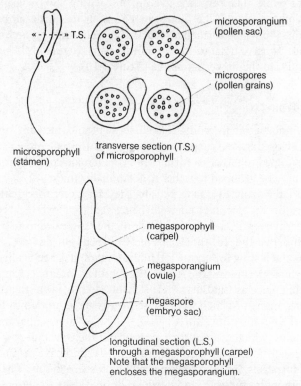

Figure 1.10   Angiosperm reproduction.   Development of the Seed Habit. In the Seed Plants the microspores and megaspores are retained on the parent plant. The microspores which contain the male sperm are transferred (by wind or by insects) to the megasporangium where a sperm fertilizes the female ovum within the megaspore. The resulting zygote divides to form an embryo within the megasporangium, the whole being known as a *seed*. In the gymnosperms the seed is not enclosed in the megasporophyll. In the angiosperms the seed is enclosed by the megasporophyll to form a fruit.

present day and have nine genera scattered in sub-tropical and tropical latitudes. It is possible that, having outgrown their evolutionary usefulness, they are now heading for extinction.

This is not, however, true of the main branch of the gymnosperms, the Coniferales, for, although at least one of its families is extinct, and some may be declining, others are still in course of active evolutionary development, e.g. there are some 90 species of pine, forty species of *Abies* and *Picea*, and ten species of larch; many are of considerable economic importance as sources of timber and pulpwood. As a group, the conifers are today among the dominant forest trees of the world in the Northern Hemisphere. *Sequoiadendron giganteum* of California is probably our biggest (some specimens are over ten metres in diameter and ninety metres in height) and longest-living organism (some may be over 4000 years old).

The reptiles

The reptiles, which made their first appearance in the Carboniferous Period of the Palaeozoic Era, were to proliferate and populate the earth, the water and the air during the Triassic and Jurassic Periods of the Mesozoic Era. At the end of this spell of dominance, most disappeared during the Cretaceous Period, giving way to egg-laying and marsupial mammals.

Reptiles in the water and in the air

Some of the massive reptiles of this remarkable time developed in the water. The plesiosaurs, which have been described as being like "long snakes stuck through the body of a turtle", had limbs like paddles; the ichthyosaurs were fish-like in shape with no neck, and the limbs were like paired fins.

Another group became adapted for flight—the pterosaurs had membranous wings which were supported by the fourth finger of the front limb. The first three fingers had little claws by means of which the animal could cling to the rocks. Their bones were hollow, thus ensuring lightness for flight. Within this group, species varied in size from a few inches to some that had a wingspan of 27 feet.

The most fascinating group of all the reptiles were the dinosaurs, for within them the genetic material exploded into a bizarre variety of shapes and sizes. If we did not have fossils to prove their existence, it would be easy to dismiss them as figments of Man's imagination. *Tyrannosaurus* was the largest carnivorous land animal that has ever lived. It weighed

10 tons, was 50 ft long and, standing on its hind legs, it was more than 20 feet tall. Its forelegs had huge hooked claws, and its large hinged lower jaw gave its face a perpetual gape. It had rows of pointed teeth, some of which were more than a foot long.

Not all dinosaurs were carnivorous. With its long neck and tail, *Brontosaurus* attained a length of about 80 feet, weighed 40 tons, and was exclusively a herbivore. It had a very small brain encased in a very small head. It is clear that its legs were not strong enough to support the heavy body, and this creature must have lived in the swamps. *Stegosaurus*, the plated dinosaur, was about 20 feet long, had a double row of projecting plates down its back, and spike-like spines on its tail. Its brain could have been no bigger than a walnut.

These and other dinosaurs dominated the earth for more than 160 million years, and then they became extinct. No-one knows why. Certainly the climate was becoming colder, and the dinosaurs may have had little ability to control their body temperature. Today's reptiles, if they are to survive in colder climes, must hibernate. Possibly the dinosaurs did not have the hibernating mechanism, and so they died. On the other hand, a strong case (Dr. T. Bakker, *Scientific American*, April, 1975) has been made that the dinosaurs may have been warm blooded and that the reduction in numbers was due to the draining of the shallow seas on the continents and a lull in mountain building, producing vast monotonous plains which, decreasing the variety of habitats, led to the collapse of the intricate ecosystem within which the dinosaurs had previously thriven. They did not become extinct, however, says Dr. Bakker. They evolved into birds.

Possibly many had become so highly specialized that they were unable to adapt to changing conditions. Possibly they were pushed aside in the evolutionary race by a group of animals that did not lay eggs—animals that retained their fertilized eggs within their bodies and allowed them to develop there—the mammals.

The birds

Thus by the end of the Cretaceous Period, the great reptiles had become extinct—all, that is, except one line (and it gave rise to the birds)—the "glorified reptiles".

We have already noted (page 21) that the pterosaurs developed membranous wings, but they were very limited in flight, and probably did little more than glide. There was, however, a second group of reptiles that took to the air, and they developed feathers—a modification not seen in the pterosaurs. Feathers are made of keratin, as are reptile scales, and it is

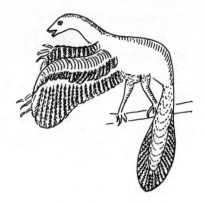

Figure 1.11   What *Archaeopteryx* might have looked like. Note large tail with feathers, also wings with claws, and teeth in mouth.

now widely believed that they are derived from such scales. They are unique to the birds, and they date back to the Jurassic Period (some 136 million years ago, see Table 1.1).

We have no fossils of links between scales and feathers, but there is no doubt that, if the fossil birds that have been found had no feathers, then we would classify them as reptiles. One of these, *Archaeopteryx* (figure 1.11), was found in Bavaria in limestone rocks of the Jurassic Period. The find was an exciting one because the rudimentary wings had claws, there were teeth in the mouth, and the bird had a long tail with a row of feathers along each side of the slender chain of vertebrae. Possibly this early "bird" could not fly. An efficient bird has to be light and powerful, and much of the lightness is achieved by hollow fragile bones. It is therefore perhaps not surprising that the fossil evidence is thin and confined largely to sea birds preserved in the chalk deposits. As well as ensuring lightness, feathers also formed a very good insulating cover for the body, and this rather than flight might have been their primary purpose.

As we have noted, the ancient reptiles may have had little control over their body temperatures, which rose and fell in step with the temperature of the air or the water. One of the major achievements of the birds or their forerunners was that of a fixed body temperature which permitted a high energy output, no matter what the external temperatures. Feathers helped to maintain it.

One of the main reasons for the success of the birds was that they moved into a new environment (the air) where they had little competition and few enemies. Here they evolved into the myriad colourful forms that we know today. Smell is of little importance to birds, and most of them have lost (or not developed) this faculty. Sight, on the other hand,

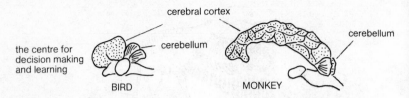

Figure 1.12   A comparison of bird and monkey brains. Note massive development of the cerebral cortex in the mammal, whereas the cerebellum (which is the centre for balance and posture) does not differ much from that of birds. The cerebral cortex is further developed in humans.

is of paramount importance, and this faculty was developed to a very high degree. Although birds were well established by the end of the Mesozoic, it was in the Tertiary Period (beginning about 70 million years ago) that they underwent their population explosion, ranging in size from the humming bird to the ostrich; in colour from the crow to the bird of paradise.

We noted earlier (figure 1.6) that the reptiles developed two atria of the heart; the birds took this a step further by also separating the ventricle into two. Thus the arterial blood was kept from mixing with the venous blood; this led to much greater efficiency and increased energy output.

The brain evolved too (figure 1.12). A good brain is required for the complications of flying, and the required parts of the brain were selected by the environment for this. Other parts, e.g. those concerned with problem-solving, remained undeveloped—birds solve their problems by flying away from them! In this connection, environmental problems are overcome by some birds by migration. One very interesting theory about the origin of migration suggests that birds originated when all the continents were together in the land mass known as Pangea (see below). Birds attempted to reach their homeland to breed each summer as the continents very slowly drifted apart, and this process has continued every year since then, young birds learning the route by being guided by the old.

One feature of birds that had great evolutionary potential was that they looked after and hatched their eggs, and subsequently fed the young. Reptiles, on the other hand, left their eggs (and the subsequent young) to fend for themselves, and were consequently that bit more vulnerable.

In the Mesozoic Era, fascinating changes were also taking place in the structure and shape of the land. The original land was beginning to break up and to separate into individual continents by the process of Continental Drift.

One of the major factors that has led to the biological diversity of the environment has been the movement of continents. Over sixty years ago, a German scientist, Alfred Wegener, put forward a proposal that at one time all of the land on earth had formed one mass, which he named Pangea. Pangea eventually split, and the present-day continents were formed as the several pieces started to drift apart. His views were not accepted, and it is only in comparatively recent times, as a result of investigation by sophisticated instrumentation, that Wegener's theory has been shown almost beyond dispute to be true, and that the process is a continuing one. Studies of magnetic fields show that the only matching possible is by the acceptance of Wegener's theory, and that the Atlantic Ocean is opening at the rate of 1–2 inches per year. Further evidence has come from computer studies on the "fit" of various coastlines based on their continental slopes. The rock strata on the west coast of Africa match with those on the east coast of South America, even gold and tin layers lining up in a quite remarkable way. The fossil record is equally important. Biologists in the nineteenth and early twentieth centuries noted the differences and

Figure 1.13 Fit of continents.

resemblances between fossils on different continents and generally agreed that links had existed between, for example, Africa and Brazil until early in the Cretaceous, some 140 million years ago, but it was generally accepted that these links were "land bridges" that had spanned the distances between the continents. These bridges, according to the theory, had subsequently sunk without trace. It has now been shown fairly conclusively that such bridges never existed. In the modern theory the continents are carried as passengers on large rigid plates which are forced apart where material wells up to form new sea floor.

It seems that the break-up of Pangea (figure 1.13) began about 200 million years ago, and this initial separation into Laurasia (North America, Europe, Asia) and Gondwana (Africa, Antarctica, Australia and India) required some 20 million years of drift. By the end of the Triassic (180 million years ago, see Table 1.1), the Australia/Antarctica mass had begun to separate from the South America/Africa group, with India beginning a long journey northwards. During this period the Atlantic and the Indian Oceans were formed.

This separation of continents led to what is termed "genetic isolation", so that each separated land mass evolved its own individual plants and animals which, because of intervening water masses, could not be "diluted" or eliminated by plants and animals from other continents.

Angiosperms (angos = a vessel, sperm = a seed)

The Mesozoic also saw the beginning of the angiosperms that were to become the dominant plants on earth. At the present time comprising some 300 families, they include some 275 000 species and they are world-wide in their distribution.

As we have seen, during the Cretaceous Period a great "dying out" occurred all over the world—the dinosaurs, the Bennettitales, many of the cycads, some of the conifers. During this same period the angiosperms arose. Why were they so successful? It would appear that a combination of features made them so, including the development of:

(1) Much more efficient water-conducting systems (xylem vessels)
(2) Broad leaves which had a greater capacity for photosynthesis
(3) A very efficient system (sieve tubes) for transporting the products of photosynthesis to all parts of the plant
(4) The ability to shed their leaves which, though such an asset in summer for photosynthesis, were a liability in winter, causing excessive water loss
(5) Tap root systems that were able to reach sources of water deep in the earth, and also able to anchor the plant in the soil, thus withstanding severe storms
(6) A more effective reproductive system—some 2–4 times more rapid than that of conifers.

It appears that in time some of these angiosperm trees lost their capacity for secondary growth, thus giving rise to a very productive line of herb-aceous types that clothed the Earth. Some of these lived for only a single season, set seed and died; others shortened their lifespan to only a few months, weeks or, in the case of some desert plants, days. The capacity for variability was enormous. Some developed underground organs, adapting their roots, stems and leaves so that the above-ground parts could die down when conditions were unfavourable and reappear when conditions improved.

A special feature of the angiosperms (distinguishing them from the gymnosperms) is that the seeds are enclosed by the leaf on which they are borne—the carpel (figure 1.10). Within or external to this there may be considerable development to form what is known botanically as a *fruit*, with the seed or seeds enclosed within it. The fruit may be succulent, e.g. cherry, blackcurrant, or may remain dry, e.g. sycamore, elm.

One other important feature which led to the dominance of the angio-sperms was the development of the flower. A few angiosperms have got very reduced flower structure, having disposed of all but the essential reproductive organs, but in most the flower consists of leaves which have been modified for protection of (sepals) and attraction to (petals) the reproductive structures (pollen grains or microspores) which are de-veloped within the stamens, and the embryo sacs or megaspores which are protected within the carpels).

Insects play a major part in the transference of the pollen grains to the carpels, and many and varied are the devices that angiosperms have evolved in order to attract them to the flower—nectar, perfume, colour, honey-guides. The flower itself has undergone considerable morpho-logical modification in order to bring about effective transfer of the pollen.

In general, angiosperms that are wind-pollinated (e.g. many grasses) are marked by rather negative characteristics—they lack nectar, perfume, or highly coloured petals. It is thought that the first angiosperms were probably wind-pollinated, and they may have had occasional visits from insects which used their pollen as food. It is certainly true that such insects as there were in the Carboniferous Period had biting jaws, and it is not until the Tertiary Period, when there was a massive explosion of flowering plants, that most of the modern families of insects also appeared. The rise of the flowering plants coincided with the rise of the insects, par-ticularly in the Cretaceous Period which began about 135 million years ago. There are good fossil flowers and seeds from this period. Flowering plants and insects owe their prominent position to each other, most plants being dependent on insects for pollination, and many insects being de-pendent on plants for food. The number of individual insects alive at any one time is said to be in the region of 10^{18}. The majority of them are either innocuous or helpful so far as man is concerned. An important minority, however, are harmful in that they compete with man for his food and are responsible for the transmission of many human and animal diseases (see Vol. 3, *Health and the Environment*).

The angiosperms are divided into two main classes based on the number of seed leaves. The dicotyledons (e.g. buttercup, tomato, rose) have two seed leaves, and the floral parts are generally in fours or fives. The members of the monocotyledons (e.g. grasses, cereals, orchids) have only one seed leaf, and the flower parts are generally in threes or multiples of three. The monocotyledons probably evolved from primitive dicotyledons. In time the monocotyledonous grasses were to clothe great areas of the world—the prairies of North America, the pampas of South America, the steppes of Russia, the grasslands of Australia and New Zealand. They provided feeding grounds for the animals that man was later to domesticate. From the grasses also sprang the cereals that were to be developed by man to provide his basic foodstuffs. Indeed the development of mankind has only been possible because of the development of the monocotyledons.

Mammals

The characteristic features of mammals are warm-bloodedness, hair which serves to conserve the body heat, and mammary glands by means of which the young are suckled by the mother. Another striking feature is the remarkable development of the brain. A. S. Romer (see page 31) has said that "on average, any mammal is an intellectual giant compared with most other vertebrates". This is due to the massive development of the paired cerebral hemispheres (see figure 1.12) These are relatively small in the amphibians and the reptiles, but in the mammals great sheets of nervous tissue, the so-called *grey matter*, developed over the expanded lobes to form what is called the *cerebral cortex* which is made up of millions of nerve cells all interconnecting with each other and with the other parts of the brain. The highest mental faculties of a mammal are located here. It is the centre of decision making and it is the centre of learning. Its presence means that its owner can store experiences, is capable of learning and of being trained. Within this cortex lies the secret of the success of the mammals. As we go up the evolutionary scale in the mammals we find that there is a corresponding increase in complexity and folding of the cortex, so that in man it completely covers the brain structures from which it was developed.

The first mammals that can be clearly identified are from the Jurassic rocks (around 165 million years ago)—small rat-like insect-eating creatures.

(a) The egg-laying mammals

The early forms were probably egg-laying, and in this they reveal their reptilian ancestry; it is exciting that we have got some representatives of this group still thriving today. The duck-billed platypus and the spiny anteater escaped extinction when they were cut off in the continent of Australia, and thus escaped intense competition from the rapidly-evolving more-advanced placental mammalian that developed on the great land masses of Europe and Asia.

Probably in very primitive mammals the mammary glands had not fully developed and the milk, as in the duck-billed platypus today, was secreted from glands in the belly which oozed out through slits and was licked by the young from the fur.

(b) The marsupial mammals

A second group of mammals became more highly evolved. Instead of laying eggs, they retained their young within the reproductive tract of the mother. There the young secured some nourishment from the yolk sac for a brief period before being born at an embryonic stage of development. On being born, they made a hazardous climb into a pouch, the marsupium, on the mother's abdomen, where they attached themselves to the nipples of the mammalian glands where development was completed. We are extremely fortunate in that some examples of these marsupials, e.g. kangaroos and opposums, are to be found in Australia. From the fossil evidence it would appear that the marsupials were highly successful, but in time they were unable to compete with the more-advanced placental mammals. One of the last places in addition to Australia, on which they retained a hold, was South America (which at this period (Cretaceous) was cut off from North America). No true placental mammals evolved there, and in their absence a series of marsupial wolf-like and cat-like forms developed. When connections were re-established between the North and South, these marsupials failed to survive the competition from the placental forms that flooded in from the North, and many became extinct.

(c) The placental mammals

Marsupial mammals and placental mammals differ in one major respect. The young of the former are born as embryos. The young of the latter are retained within the mother to an advanced stage, and they are fed through a special organ—the placenta.

This development has been very successful; placental mammals have spread all over the Earth, replacing the marsupials wherever they have been in competition. The mammals came to dominate the Cainozoic Era as the reptiles had dominated the Mesozoic. Some 60–70 million years ago there was a massive "explosion" of mammalian types. By the middle of the Cainozoic, mammals were thriving everywhere, grazing, swimming, hunting, flying. They expanded to fill the gaps, the "ecological niches" that had been left by the vanished reptiles; warm climates were widespread and the mammals luxuriated in them.

They ranged in size from shrews weighing a few ounces to whales weighing many tons; from grass-eaters to carnivores. And they have remained dominant to the present day. How do we account for their success? Quite simply, it appears, because they retained their embryos within their bodies and looked after their young. In this there is a very striking parallel with the angiosperms (the most successful group of plants) which, instead of throwing their spores to the winds, held them in their bodies and, after fertilization, retained their embryos (as seed) to a fairly advanced stage of development.

The message seems plain. For evolutionary success, protect the embryo.

Conclusion

And so after 5000 million years we reach the present day. Our atmosphere is composed of 78% nitrogen and 21% oxygen, with the remaining 1% made up of carbon dioxide, hydrogen, and a few heavy gases. The amount of water vapour varies from region to region. We are protected from the harmful radiations (that, rather anomalously, once brought living things into being) by an ozone layer, and we are kept temperate by a cloud layer. Our planet Earth has a 30% land area and 70% water area—six great oceans and seven great continents. It has an irregular surface, ranging from 29 028 feet above sea level (Mount Everest) to 36 198 feet below sea level (the Marianas Trench off the Philippines in the Pacific Ocean).

Living things are found almost everywhere, ranging from the microscopic plants and animals to the giant trees of California and the whales of our oceans. There are living things that last only a few hours, and some that have a life span of thousands of years.

In the seas there are blue-green, green, red and brown algae, invertebrates ranging from sponges to squids, fish and a few mammals that have found their way back from the land.

In the air there are birds with a great array of plumage and a great range of song. Insects share this medium but, not content with it alone, they spill over on to the land, which is covered by their symbiotic partners, the angiosperms. Placental mammals are everywhere, including the most successful placental mammal of all—man.

We are fortunate in that we still have living reminders of the past in our mosses, liverworts, ferns, cycads, gymnosperms, coelacanths, amphibians, reptiles, egg-laying and marsupial mammals. We must cherish these. While man's effect on the environment (the subject matter of this whole series of books) is clearly outside the scope of this chapter, we could note that in the past few hundred years, with the advance of technology, the life of man has been improved out of all recognition. Great care must be taken that in improving his own lot man does not also destroy his natural heritage.

FURTHER READING

Fletcher, W. W. (1974), *Modern Man Looks at Evolution*, Fontana/Collins.

Fletcher, W. W. (1976), "Our Environment—past, present, future," *British Pharmaceutical Journal*, Oct. 9.

Hallam, A. (1972), "Continental Drift and the Fossil Record," *Scientific American*, Nov.

Hallam, A. (1975), "Alfred Wegener and the Hypothesis of Continental Drift," *Scientific American*, Feb.

Jensen, W. A. and Salisbury, F. B. (1972), *Botany—an Ecological Approach*, Wadsworth.

Romer, A. S. (1968), *The Procession of Life*, Weidenfeld and Nicolson.

Scagel, R. F. *et al.* (1965), *An Evolutionary Survey of the Plant Kingdom*, Blackie.

Scagel, R. F. *et al.* (1969), *Plant Diversity: An Evolutionary Approach*, Wadsworth Publishing Co.

Stansfield, W. D. (1977), *The Science of Evolution*, Macmillan.

CHAPTER TWO

HUMAN EVOLUTION AND THE ENVIRONMENT

G. AINSWORTH HARRISON

THE ENVIRONMENT ACTS ON MAN, AS ON ALL OTHER ORGANISMS, IN TWO distinct but interrelated ways: first, in moulding the phenotype (total characteristics) in individual development through interaction with the individual's genetic constitution; and, secondly, in favouring certain phenotypes in survival and reproduction. It is the latter form of environmental action which is the main evolutionary force, since it constitutes natural selection when there is a genetic component to the differences between phenotypes. Natural selection will often, though not invariably, cause a systematic change over time in the frequency of genes in a population, and this is the essential feature of biological evolution. The role of the environment in development cannot, however, be totally ignored, since natural selection of the form classically conceived by Darwin, does not act directly on genes but on the whole phenotypes of individuals—and all phenotypes are products of the environment in which they develop and live. The effect of natural selection is to fit or adapt organisms increasingly to their environment, and we can think of it as moulding the gene pool of a population to an environment in such a way that, as the genetic constitutions arise through reproduction from the pool and interact in development with that environment, they produce individuals who are adapted to their environment. In this chapter attention will be primarily directed to considering the ways in which man's environments have acted as selective agents, but the effect of environmental factors in ontogeny must always be remembered. While environments can and do change independently of the organisms which inhabit them, it is worth remembering that biological evolution itself necessarily also produces environmental change. And this is true not only for different species but also for different members of the same species which are in competition with one another. Thus the origin of a new variety, better adapted to the prevailing environment than the old variety, will often not only change the environment for other co-habiting species but also for its own species and indeed for itself. Evolution can thus in some ways be thought of as always chasing the environments it itself is producing. This general phenomenon is particularly relevant in considering the human situation.

In practice, evolution is examined on two scales; the macroscale, as evidenced in the comparative anatomy and physiology of varyingly related forms and, particularly, by the fossil record indicating the descent of these forms; and the microscale, as evidenced in the genetic variety and its significance in contemporary populations. Considerable information is now available on both scales for interpreting human evolution and examining at least some of the environmental selective factors which have operated to produce man as we now find him. This chapter will be devoted mainly to macroscale phenomena, and the following chapter to microscale phenomena.

The primates

Man belongs to the mammalian Order of Primates and is most closely related among contemporary forms to the lemurs, lorises, tarsiers, New and Old World Monkeys and, especially, the Great Apes. All these primates have had an arboreal ancestry and many of them, of course, still live primarily in trees. Most of the features they share in common can be seen as adaptation to this type of environment. Development of the hand and/or foot as a grasping organ, replacement of sharp claws by nails, and a wide range of movement of the forelimb are features which facilitate living in the three-dimensional irregular and precarious environment of the trees. Similarly, general reduction in the sense of smell, but emphasis on visual and auditory acuity, and well-developed tactile and balancing senses are concomitants of the arboreal habit. These changes in the senses are at least partly responsible for the fact that all primates tend to have relatively large brains. Even the tendency for primates to have only a small number of offspring at any one time can also be seen as an arboreal adaptation, for parental care of the young is at a premium in the trees. An arboreal environment is one of the few in which a reduction in fertility may improve an animal's reproductive success. And a small litter size has important consequences, since it means that there is likely to be less competition between litter mates, and therefore no special advantage in rapid development. Primates are characterized by having a long period of development in their life histories. All these features man shares, and with hindsight, they can be seen to have been critical in his emergence as the dominant mammalian species. As compared with other primates, the distinctive feature about human evolution is that man's ancestors abandoned an arboreal environment and became terrestrial at just the right stage of their evolutionary history.

Bipedalism

Other primates like the baboons and their relatives and the gorilla have also secondarily adopted a primarily terrestrial life, but none have achieved the same evolutionary dominance as man. The critical feature of human differention is that man's ancestors, the early hominids, in descending from the trees adopted an erect posture and bipedal loco-motion. The effect of this development, superimposed on the inherited arboreal adaptations, turned out to have dramatic consequences. It is not that bipedalism is a particularly effective mode of locomotion, but it does emancipate the forelimb from all locomotor function, and allows it to be used for other things. Thus the hand would be employed, first for using natural objects as tools, and later for making such tools. This capacity was, no doubt, immediately advantageous when man's ancestors, with few intrinsic mechanisms of defence, moved into an environment with a much larger number of predators than are found in the trees. However, it was also the beginning of an important interrelated evolution of hand and brain which led both to the present human hand with its fine tactile sensitivity and motor dexterity, and the present brain in which there is such extensive representation of the hand in the sensory and motor cortex. And the tools which no doubt were initially used mainly in defence became refined as instruments of offence. Most primates are mainly vegetarian and the first hominids were probably likewise. There is evidence, however, in the fossil record, that at least by the late Pliocene some hominid groups had become omnivorous if not strictly carnivorous. The form known as *Australopithecus robustus* (or *Paranthropus robustus*) was almost certainly vegetarian, as evidenced by the structure of its teeth, but the contemporary *Australopithecus africanus* was probably a hunter-gatherer.

The freeing of the forelimb from support and locomotor function also allowed it to be developed for carrying things. This facilitated the establishment of a home base. It also meant that a baby can be brought into much closer and more intimate contact with its mother than is possible in most mammals. The human baby is born in a very immature state—partly because of the limitations imposed on the female pelvis by bipedalism—and is long dependent upon intensive maternal care. It has been suggested that the unique feature of human sexual behaviour, in which the female is receptive to the male through most of her menstrual cycle, is to be seen as an adaptation to long infant dependence and a hunting mode of life in which a mother could not fully participate while caring for her child. Her constant sexual receptivity, it is surmised, tied a particular male to her to provide food and protection.

Thus the consequences of bipedalism were ramifying, and the fossil record indicates that it was an early feature of hominids, arising before the great reduction in jaw size and expansion in brain size which also characterizes the hominid lineage. Just why man's ancestors first became terrestrial, and in doing so became bipedal, is unclear. The divergence of apes and man probably occurred in the early Miocene, though there is no fossil record of this. At that time, however, tropical Africa, in which the hominids almost certainly differentiated, was experiencing a reduction of forests and an expansion of grasslands. It seems likely that it was the food opportunities offered by these grasslands that caused a group of formerly tree-living primates to move into them, and there is some suggestion that the earliest fossil hominid known, *Ramapithecus*, which has been recovered from deposits in India and Europe as well as in Africa, was adapted to a grain diet. The adoption of bipedalism may have been facilitated by the immediate ancestors being semi-brachiators, like present-day langur monkeys. This form of locomotion, in which the animal periodically swings by its arms through the trees, puts the body into an erect posture and requires adaptations to this posture. It has also been suggested that there may have been advantage in standing on the rear legs to see over the tall grasses, as the first hominids were small animals.

Adaptation to temperature

As already indicated, these animals were tropical, and this in itself may account in part for the fact that man's tolerance of high temperature is good. This tolerance is, however, exceptional and would seem to exceed that of most other tropical animals. It is achieved through man's naked-ness and the evolution of an unmatched sweating capacity through eccrine glands. The constant flow of water over a richly vascular naked skin permits a rapid dissipation of body heat by evaporation in hot dry environments such as occur on the savanna.

The system possibly evolved in response to man becoming a day hunter. Most of the tropical carnivores hunt in the relative cool of the early morning or evening, leaving a vacant ecological niche in the heat of the day. And early man without great speed or weaponry for rapid "kills" would probably have had to track wounded game for many hours. This certainly is the experience of some modern hunter-gatherers such as the Kalahari Bushmen who practice an economy which may not be very different from that which prevailed throughout much of human evolution. The Bushmen are exposed to very high heat loads in both their hunting and gathering activities, particularly at certain times of the year.

While one naturally thinks of the heat as the dominating climatic factor in the tropics, it needs to be remembered that the savanna, and even more the neighbouring deserts, experience marked diurnal temperature variation and the nights can be very cold. Some of the capacity for all members of the present human species to adapt physiologically to the cold may be explained through this. However, it seems more than likely that hominids were not able to move into temperate and cold climatic regimes until they were able to provide themselves with some technological assistance. Fossils of *Ramapithecus* and the australopithecines are confined to regions of the world which are hot, or were at the time of their existence. The genus *Homo*, as evidenced from recent palaeontological finds in East Africa, probably also arose in the tropics as a branching of the *Australopithecus africanus* lineage, and the first-discovered remains of *Homo erectus* come from Java—another tropical environment. But *H. erectus* remains of the Lower and Middle Pleistocene are also known from Europe, North Africa and North China, in environments that were temperate. At Chou-kou-tien, near Peking, were found the human fossils associated with the bones of the now extinct species of bison, horse, rhinoceros, flat-antlered deer, brown bear, big-horned sheep, mammoth, camel, ostrich, antelope, water buffalo, wild boar and hyena.

Tools

There is some dispute as to whether the australopithecines made, as distinct from used, tools. The problem is in part a semantic one, depending upon which fossils are assigned to *Australopithecus* and which to *Homo*. In the continuum of gradual change which represents evolution, the discrete categories of taxa into which organisms are classified are bound to be somewhat artificial. But there is some evidence from the various caves in South Africa, from which many australopithecine fossils have been recovered, that these forms used, and partially fashioned, bits of bone, teeth and horn as tools—the so-called "osteodontokeratic" culture—and in East Africa some of the fossils are associated with simple stone tools which have been clearly manufactured. In the earliest bed at Olduvai Gorge there is evidence of a living site which was probably occupied by a form variously known as *Homo habilis*, *Australopithecus habilis* or *A. africanus habilis*. By the stage of definitive *H. erectus*, quite refined stone tools were being manufactured, and at Peking there is evidence of the use and probably making of fire. Here were found charred fragments of fox, large cat, macaque and baboon. Interestingly the wood ash reveals the existence of a dwarf species of hackberry whose fruit is rich in sugars and vitimin C. Although probably of little nutritional significance, there is also evidence of cannibalism in the human remains from Chou-kou-tien, in that the base of the skull is missing from all the specimens.

Differentiation of *Homo*

With fire, the use of animal skins as clothing, and simple habitations such as caves, *H. erectus* was able to colonize the temperate, if not the really cold, ecosystems of the Earth, and as already indicated, there is evidence that the species was widely distributed in Africa and Eurasia. Anatomically *H. erectus* differs from *Australopithecus* in having a larger brain (with a cranial capacity of around 1000 cm^3 as compared with 500 or 600 cm^3), smaller jaws (which could in part be due to the cooking of food), and perfection of the bipedal adaptations (the post-cranial skeleton is indistinguishable from modern man).

Most authorities recognize that *H. erectus* was the direct and immediate ancestor of the present species *H. sapiens*. The evolutionary change mainly involves further increase in brain size to around 1400 or 1500 cm^3, and further reduction in jaw and tooth size. The detailed pattern of the evolution is still, however, not clear. Most of the evidence comes from Europe and Africa, and that from Europe is the better in that the geological dating is firmer.

The second part of the Pleistocene in high latitudes is dominated by a series of Ice Ages, when there was extensive expansion of such ice sheets as the Arctic, Scandinavian and Alpine glaciers. Using the terminology of the Alpine glaciations, there were four main Ice Ages: Günz, Mindel, Riss and Würm, of decreasing antiquity. They were separated by three *interglacials* when climatic conditions improved and, indeed, at times became sub-tropical. Each of the major Ice Ages also experienced at least one and, in the case of the Würm, two periods of climatic amelioration, known as *interstadials*. Until the time of the last glaciation, all the fossils recovered came from interglacial or interstadial periods, and mainly the former. Only a few are known from the Günz and the Mindel period, e.g. the Heidelberg jaw, and are usually ascribed to *Homo erectus*. The later forms from the 2nd Interglacial through to the Würm I glaciation tend in skull morphology to be intermediate between *H. erectus* and present-day man, but some are quite modernistic; and this applies not only to the later specimens, but also to some early ones such as the Swanscombe remains from 2nd Interglacial river-terrace deposits in Kent and the Fontéchevade remains from the last interglacial of the Dordogne.

One of the anomalies of human evolution is that almost all these fossils appear more like ourselves than the people who occupied Europe and surrounding regions during the Würm I glaciation. The latter form the group known as Classical Neanderthal man and are characterized by having low skulls with little development of a forehead and a heavy bar of bone above the eye region. They lack a chin and have poor development of a cheek region. In these features they resemble *Homo erectus* but their brains were at least as large as those of present-day man. How it comes about that primitive features follow advanced ones has been a matter of much speculation. It has been pointed out that the Classical Neanderthalers were the first people to occupy an arctic type environment, and that much of Europe was almost totally isolated from the rest of Eurasia during the Würm glaciation, when the Scandinavian and Alpine Ice Sheets almost met in Central Europe. The features of Classical Neanderthal man have thus been seen as cold adaptations evolved in isolation and just happening to appear as a reversal of the main evolutionary trend. However, contemporaneous remains of Classical Neanderthal man have been found outside the area of European isolation, as for example in Israel, Iran and Southern Russia, and in regions which never became exceptionally cold. Further, the features of Classical Neanderthal man could not contrast more with those of Eskimo and present Arctic peoples who are well adapted to the cold.

It seems more likely that what the fossil record is representing is a two-lineage descent from *H. erectus*: a southern tropical and subtropical one evolving rapidly towards present-day man and which only occupied Europe during the warmer periods, and a northern one, evolving in the same direction but more slowly, perhaps because of smaller population sizes in the sub-arctic environments, and only occurring in Europe when conditions there became cold. Some fragmentary remains from the Riss Interstadial as well as Classical Neanderthal from the Würm I period would support this view. It further needs to be remembered that much of central Asia has been poorly explored for fossils. The presence of two lineages of descent from *H. erectus* is not exceptional, and indeed the occurrence of Rhodesian man in Africa, and Solo man from Upper Pleistocene deposits in Java, suggests that there may have been more. Nor does the existence of lineages imply their total isolation, and human taxonomists nowadays take the view that all the forms succeeding *H. erectus*, including Classical Neanderthal man, should be grouped as *Homo sapiens*.

Modern man

It was not, however, until about 40 000 years ago that men, indistinguishable from ourselves in their skeletons, first appeared. The earliest known, of this age, come from the Niah cave in Borneo, but ones of similar antiquity have been found in Europe (e.g. at Combe Capelle and Cro-Magnon in France), East and South Africa and China. In Europe the time corresponds to the retreat of the Würm I glaciations and the opening up of migratory routes with Asia and the Middle East. It seems likely that modern man came in along these routes. The economy was still a hunting one and the tools were made of flint and stone, but these tools are quite superior to those made by Classical Neanderthal man. The latter is associated with a form of flint industry called the Mousterian. The new forms of people, the men of the Upper Palaeolithic, manufactured stone knives, arrow heads, engraving tools and other finely fashioned implements. They are also the first people who are certainly known to have had ritual burial of their dead and to have developed a refined aesthetic sense, as is evidenced in their cave art.

Man in Europe

It is therefore not surprising that they supplanted the Classical Neanderthalers and quickly came to occupy the whole of Europe. Whether or not there was intermixture is not known; there is very little, if any, evidence for it in the morphology of the skeletal remains that have been recovered, but this is not surprising if the newcomers greatly outnumbered the Neanderthalers. The implication in the taxonomy of putting both Neanderthal man and the modern type of man in the same species *H. sapiens* is that intermixture would have occurred.

   Once in Europe, the modern-type men continued to inhabit it not only through the first and second interstadials but also through the Würm II and III glaciations. There is a succession of flint industries and, during the Würm III phase, we find implements made of bone and ivory, not unlike those used by modern Eskimo. This is so-called Magdalenian industry. The human remains are, however, rather homogeneous, though again some similarity with Eskimo is evident in the skulls from the Würm III glaciation. This probably represents common adaptations to the cold rather than any direct relationship. Overall, the men of the Upper Palaeolithic were tall and robust in physique, as suited hunters in a cold environment. People of much the same form continued into the Mesolithic or Middle Stone Age, when similar life-styles to those practised during the Upper Palaeolithic were extended into the post-glacial period, but the technology and material culture became more diversified. At this time we find that many of the stone tools were small, such as fish hooks and tiny arrowheads forming microlithic industries, and there is evidence of the invention of basketry.

Man in Africa

Broadly similar events were occurring in Africa, but the absolute dating of them and their intercorrelations in different geographical regions is not as certain yet as in Europe. Most of Africa appears not to have been affected by the Ice Age succession which provides so much of the geological dating framework in Europe. It was once thought that a series of African wet phases, Pluvials, and dry periods, Interpluvials, corresponded respectively with the cold and warm periods in higher latitudes, but this correlation is now strongly questioned. Quite a number of fossil remains of modern-type man have been recovered in Africa, mainly East and South, and many of these were probably contemporaneous with the Upper Palaeolithic forms from Europe. As here, there would also seem to have been archaic forms known as Rhodesian man and Saldanha Bay man, which themselves were probably contemporaneous with the Neanderthalers and may well have been another lineage from *H. erectus.* Interestingly, none of the modern-type skulls shows typically negroid features until quite late in the sequence. It is surmised that the negroes arose first in the forest areas, where conditions for fossilization are poor, and subsequently spread throughout sub-Saharan Africa. Certainly there is strong linguistic and anthropological evidence that the Bantu-speaking groups, who now occupy most of Eastern and Southern Africa, differentiated in a quite small area on the northern forest/savanna fringe and subsequently spread to the other regions. These migrations, however, were very late and were still occurring in South Africa at the time of the European colonization. The earlier skulls show their closest affinity with those from recent European populations, particularly Mediterranean groups, but some of the remains, e.g. those found at Boskop in South Africa, show similarity with modern Bushmen, but are noticeably larger.

Man in Australia

Evidence for man in Australia first appears about 23 000 years ago, at Lake Mungo in New South Wales. It seems likely that the continent was occupied by population movement through the Indonesian islands and New Guinea, and a few fossils are known from this region. All the remains in Australia itself clearly belong to the modern form, though the earliest are not as obviously Australian aborigines as are later ones.

Man in the Americas

The Americas were also first colonized by hominids late in the evolutionary sequence, and all the fossils recovered are of modern-type man. It would appear that the colonization took place in a series of waves across the Bering Straits during the second half of the last glacial periods, with the ancestors of the modern Eskimos being the last to arrive. Then followed a rapid spread throughout North and South America. One of the earliest fossils comes from Tepexpan, Mexico, with a chronometric radioactive carbon date of 10 000 years. This and most of the other skulls known show affinity with present-day Amerindians.

Hunter-gatherers

The peoples of the Palaeolithic and Mesolithic—those which we have so far mainly considered—all practised a hunter-gatherer economy. This essentially involves exploiting the resources that are offered by the natural environment, but particularly in hunting required a considerable technology. It is a form of economy still practised by a number of groups, such as the Australian Aborigines, the Kalahari Bushmen, the Eskimo, the Pygmy and various Amerindians. In some, the emphasis is on hunting, as in the case of the Eskimo and Pygmy; in others, it is more on gathering, e.g. the Aborigine and Bushmen. There are difficulties in extrapolating from the lifestyles of these peoples to those which generally prevailed throughout the Palaeolithic, mainly because the hunter-gatherer economy has only persisted in refuge and hostile environments. There is also considerable variation among the present-day groups, which partly depend upon local environmental circumstances. Thus, for example, the population density is greater among Bushmen than among Aborigines, because of the presence in Africa of large game which does not occur in Australia. In southern California there happens to be a local agave plant rich in carbohydrate which helps to support an even larger population in another desert region. But certain broad features of population structure and lifestyle seem to be inevitable consequences of hunting and gathering, and it is important to consider them since they are the ones that prevailed throughout almost the whole of human evolution.

Restriction of population size

Perhaps the most important consequence of hunter-gathering is the restriction it imposes on population size. Although, as already seen, there may be marked local variation, all present hunter-gatherer groups are small, and the archaeological evidence indicates that this was always so. Throughout the Palaeolithic, the social and economic groups probably rarely exceeded a few hundred individuals and were usually much smaller. This has important genetic and ecological consequences. In small populations the frequency of genes can change markedly from generation to generation solely by chance—a phenomenon known as *genetic drift*. This particularly happens if the alternative genes present have little or no effect on fitness. In other words, evolution can occur as a stochastic process, without the action of natural selection. In ecological terms, small population size can profoundly affect the infectious-disease pattern of a population. A disease such as measles, which has no alternative host to man and produces life-long immunity in those which have been affected, requires access to about half a million people to maintain itself. The total world population of mankind was probably not much bigger than this in the Upper Palaeolithic, and smaller in earlier times. The small semi-isolated groups of hunter-gatherers today are free of many bacterial and virus diseases unless they are brought into contact with the outside world, when of course epidemics can cause catastrophe.

Other ecological features of ancestral hunter-gatherers are nutritional. Although the amount of animal protein consumed probably varied considerably, all groups are likely to have had larger amounts of cellulose and less of sodium chloride than modern diets. Periods of plenty would frequently be followed by starvation conditions. Babies, of course, were exclusively fed on human milk and, extrapolating from present-day groups, were probably not weaned until 2 or 3 years of age. Because lactation tends to suppress ovulation, this practice affects birth spacing, which may well have been important in infant survival, and reduces total fertility.

Palaeolithic life

In the Palaeolithic, there was no artificial light, no high-intensity noise, no synthetic chemicals to contaminate and pollute food and air. On the other hand, infant death rates from accident, food shortage, and worm and bacterial parasitism, were high and life expectancy low. Time budgets were probably very different from today, with feeding and sleeping more dictated by condition than by time. Amongst present-day hunter-gatherers only a small proportion of the day is spent in food collection; the economy is a low-effort one. Socially, the people one meets are people one knows; they have prescribed bonds and behaviours dictated by kinship and social rituals.

Neolithic life

All this began to change with the onset of the Neolithic. The period gets its name from the fact that at this time stone tools began to be polished, but the change of real significance was the beginning of the cultivation of plants, and the domestication of animals for food and clothing. Pottery was invented for the cooking and storage of food. Fixed settlement became possible, and desirable with agriculture and, whether as cause or effect, populations began to grow. The Neolithic started in the Middle East, just a few thousand years B.C. and spread to affect almost all the societies of the Old World. Based upon different crops and animals, it arose independently, somewhat later in the New World in Central America and the Andean altiplano. Following quickly on the beginnings of plant cultivation and animal breeding in the Old World, which seems to have arisen in the "fertile crescent" of Syria, Iraq and Iran, the societies of Neolithic revolution tended to move into the river valley systems, such as the Euphrates, Tigris and Nile, and took advantage of the rich soil and its naturally replenished fertility. Here urbanization began, and produced for the first time heavy crowdings of people, marked occupational diversity and a class/caste structured society.

In the spread of the Neolithic from these early centres it is not always clear how far this was brought about by the movement of people and how far by the movement of ideas, which can, of course, occur much more rapidly. But in either event it would appear, at least in Europe, to be associated with a reduction in human body size. The peoples of the Upper Palaeolithic were tall and physically robust; the early farmers and pastoralists were smaller and more gracile in physique. The change may be due, either through selection or developmental modification, to the fundamental changes in nutritional practice. Whereas hunter-gatherers are vulnerable to acute food shortages, subsistence agriculturalists tend to be exposed to chronic malnutrition. Under these latter circumstances, small body size is almost certainly advantageous.

Use of metals

Discovery of the use of metals, first copper and bronze, and later iron, followed the Neolithic and is associated with further population growth and movement. The spread of the Bantu-speaking peoples in Africa already mentioned appears to be associated with the possession of iron. This metal, hard and capable of holding a sharp edge, is particularly helpful in cutting down forests for agricultural land. It is an important component of slash and burn or swidden agriculture, still so widespread among traditional societies in the tropics today, though interestingly iron was unknown in the New World and New Guinea, centres of swidden agriculture, before their discovery by Europeans.

Man and the environment

With increasing technological capacity, man began progressively to change and control the natural environments and thus to impose his own selective pressures, not only on other organisms but also on himself. Many of the changes were beneficial in the sense that they promoted fertility and/or reduced mortality, but often they carried with them unforeseen and frequently disadvantageous consequences, both in the short and long term. It is doubtful whether any of the economies practised by human societies, other than hunter-gathering, is in anything like balance, and at least some of the technological development which has constantly occurred since hunter-gathering can be seen as attempts to rectify the disadvantageous effects of previous change. Evolution has operated similarly. A nice example is provided by the effects of the discovery of iron in West Africa. As already mentioned, this facilitated swidden agriculture, and there is evidence of the spread of this type of economy from east to west in West Africa following the invention of iron tools in Central Africa. The economy markedly changed the natural habitat and particularly led to extensive small pool formation close to human habitation. Such pools are ideal breeding grounds for the mosquito *Anopheles gambiae*, one of the most important vectors of falciparum malaria. It would seem therefore that malaria incidence also spread with the agriculture. There are a number of genetic systems which provide some protection from malaria, at least in early life (e.g. page 61) including sickle-cell haemoglobin and haemoglobin C. The genes for these two variants occur in West Africa, and the sickle gene shows a gradient of frequency diminishing from east to west, and thus following the pattern of agricultural and malarial spread.

Adaptations to culture

Another case, where economic factors appear to have produced evolutionary change in man, concerns the production of the enzyme lactase in the digestive system. This enzyme is present in all infant mammals to break down the milk disaccharide sugar lactose into galactose and glucose, but stops being produced at weaning. However, in those human societies which have depended heavily on milk as an adult food source for many generations, lactase is found at all ages. There is strong evidence that the system is under genetic control and is the product of natural selection. It has also been noted that there is an association among different human groups between body physique and the main types of weapons used, particularly whether these are bows or spears. Apparently spears can best be used by people of linear physique with long arms, whilst bowmen are more stocky and muscular. In this case, it is not clear whether a prevailing physique determined which form of weaponry a human group chose, or whether the choice came first and determined, through natural selection, the physique; probably it was some of each. But the situation provides an interesting example of the inter-relations between culture and biology.

Geographical variety

Natural selection in recent and contemporary human populations is discussed systematically in the next chapter, but it needs to be noted here that it has played a key role in producing the patterns of geographical genetic variation which are observable today. The form of these patterns, particularly as they appear in easily visible characters such as skin colour and hair structure, led to the establishment of racial taxonomies, and groups of people were ascribed to such major categories as Caucasoid, Negroid, and Mongoloid (which were regarded as equivalent to sub-species in zoological nomenclature), and within these to minor categories such as Nordic, Alpine and Mediterranean. It is now known from study of characters like blood groups, serum proteins, haemoglobin and enzyme variants that the situation is far less clear-cut. Populations tend to differ in the frequency with which particular genes occur, rather than in presences and absences. These frequencies tend to change gradually over extended geographical distances and there is rarely close concordance between the distribution of one genetic system and that of another. Tight taxonomies are thus arbitrary and artificial.

Racial differentiation

Interbreeding between local groups has probably been a persistent feature in human evolution, and migrations over considerable distances are not new, as is witnessed by the spread of the early hominids out of the tropics or the original colonizations of the Americas. The precise pattern of geographical variety now observed must have been profoundly affected by just which populations grew and spread, and this probably had little to do with their biology. Despite these effects and the consequent complexity of human geographical variation, it is true that at least broad continental patterns emerge. The peoples of Africa south of the Sahara do tend to share many hereditary features which are different from those of other areas. The same is true of the eastern Asiatics, the Australian aborigines and the peoples of Europe. This is not surprising since, apart from distance alone, there have been at least partial barriers to human movement, such as oceans, mountains and deserts, and groups who are mainly isolated from one another will differentiate separately both by genetic drift and in adapting to local environments.

How old are these major groups? This is a controversial question which has received considerable attention by anthropologists over the years. Some have proposed separate ancestries from *H. erectus*, pointing to some geographical variation in this species which corresponds to present-day patterns. In extreme form, however, it calls for parallel evolution of unprecedented dimension. The more common view is for a monophyletic origin of all present human groups with relatively recent differentiation, i.e. within the last 40 000 years. The extreme monophyletic view seems, however, as unlikely as the polyphyletic one; it would call for some single origin of *H. sapiens* in a particular place with the new species then rapidly spreading over the rest of the world, bringing extinction to the *H. erectus* groups which hadn't quite yet reached *sapiens* status! Apart from its intrinsic unlikelihood, it is not supported by the fossil evidence since, as already seen, remains of modern-type man appear all over the Old World at much the same date.

What probably happened is that most of the *H. erectus* groups evolved gradually towards the *sapiens* state under selection pressures which were world-wide for smaller jaws and larger brains, and that the whole complex of populations was held together as a single species by constant local intermixture and periodic long-range migration. This would be compatible with the maintenance of local genetic differentiation and adaptation, and it could thus be that some of the so-called racial characters of modern man are actually older than the species *H. sapiens*. If this is what happened, then the question about the time of racial differentiation becomes practically meaningless. We can, however, say that the overall skeletal features which characterized negroid peoples are not found until quite late in the fossil record, and until then the group must have had only a limited distribution at most. Further, there is some evidence that many of the characteristics of mongoloid groups are cold adaptations, and that these were largely evolved during the latter part of the Würm glaciations in north and central Asia.

Growth and spread of types

The subsequent growth and spread of mongoloid, negroid and caucasoid peoples have clearly had a profound effect upon the genetic composition of the human species. This growth is also the one that has had the most marked effect in changing natural and cultural environments, and thus in directing the patterns of present evolution. It is crudely estimated that by the first urban phase of the Neolithic, world population was of the order of 20–40 million. By the time of the Roman Empire the figure had reached 200 million; by A.D. 1600, 500 million; and currently about 4000 million. It has been customary to regard technological invention, through its effects on nutrition and health and thus on fertility and mortality, as the main general factor responsible for this growth. However, a strong case has been made—particularly in the development of agricultural practices—for population growth as the independent variable and technology as the dependent one. In other words, peoples have been forced to invention when population growth threatened existing resources.

Again, it is unnecessary for the situation to be an "either/or" one, and both processes could operate in sequence, with a particular technical invention providing more resources than are needed for the population size which generated it. The important issue here, however, is that in the latter phases of human existence, environments have been changing as a result of human activity at an ever-increasing rate. Disadvantageous effects of these changes can, at least in principle, be met by technological and cultural adaptations which can also change and spread very rapidly, though in doing so they are likely to promote yet further environmental change. The biology, however, is left behind, as it were, because of the intrinsic inertia in the genetic system and generation time. We can thus think of present-day man as being genetically programmed mainly for an Upper Palaeolithic life style, but actually living in a totally different way. Some consider that many of the ills they see in modern societies arise from this incompatibility, but the evidence for this is poor. There are indications that the catecholamine "flight/fight" response, which presumably was advantageous to hunter-gatherers, is involved in predisposing to coronary heart disease when it is elicited in sedentary man. But many of the changes between the Palaeolithic and now have involved removing former hazards from the environment. Certainly life expectancy is now much greater, and most would regard this as a good thing.

FURTHER READING

Boyden, S. V., Ed. (1970), *The Impact of Civilization on the Biology of Man*, Australian Academy of Sciences.
Proceedings of a scientific symposium on various evolutionary and ecological aspects of human biology.

Day, M. H. (1977), *Guide to Fossil Man* (3rd Edition), Cassell, London.
A comprehensive description of all the major fossil finds.
Dobzhansky, T. (1962), *Mankind Evolving,* Yale University Press.
A superb synthesis of the evidence for the nature of human evolution.
Harrison, G. A. and Gibson, J. B., Eds (1976), *Man in Urban Environments,* Oxford University Press.
Contains various papers dealing with the effects of present-day environments on the biology of man.
Howells, W. W. (1967), *Mankind in the Making,* Penguin.
A good popular account of many aspects of physical anthropology.
Howells, W. W. (1973), *Evolution of the Genus Homo,* Addison-Wesley, Reading, Massachusetts.
An excellent account of the fossils of our own genus and of their origins.
Hulse, F. S. (1971), *The Human Species,* (2nd Edition), Random House, New York.
A sound introduction to physical anthropology.
Katz, S. H. Ed. (1975), *Biological Anthropology,* W. H. Freeman, San Francisco.
An excellent collection of readings from *Scientific American* covering all aspects of human evolution.
Le Gros Clark, W. E. (1964), *The Fossil Evidence for Human Evolution,* Chicago University Press.
Now somewhat outdated, but in many areas still the most authoritative text on the course of human evolution.
Lee, R. B. and DeVore, I., Eds (1968), *Man the Hunter,* Aldine Publishing, Chicago.
Proceedings of a scientific symposium on the ways of life of present and past hunter-gatherers.
Pilbeam, D. (1972), *The Ascent of Man,* Macmillan.
A fine introductory textbook for first year university undergraduates.
Poirier, F. E. (1977), *Fossil Evidence,* C. N. Mosby Co., Saint Louis.
An up-to-date account which includes description of some of the recently discovered human fossils from East Africa.

CHAPTER THREE

NATURAL SELECTION IN MAN

Forbes W. Robertson

Introduction

The minimum condition for effective natural selection in a population is the presence of genetic variation in fertility and/or survival to reproductive age. If this condition is met, then there may or may not be a consistent change in the genetic make-up of the population, according to whether the selection is directional or stabilizing. If it is directional, it will act in successive generations to favour fitter individuals so as to promote a change in gene frequency. The rate of change will be proportional to the intensity of the selection, the relative importance of heredity and environment in determining the individual's rating for a selected trait, the level of variation between individuals, and also the manner in which the genes interact to influence the degree of resemblance between parent and offspring. Whether or not selection for higher survival or fertility also leads to obvious change in appearance will depend on how far the genes concerned influence both fitness and form. This sort of selection represents the classic process which promotes organic diversity by ensuring adaptation to different environments. But natural selection may also act to maintain the *status quo* when the fitness of an individual decreases as his rating, for a given trait, deviates from the average rating for the population. It is believed that such stabilizing selection has played an important part in establishing the average value for those traits which vary about an intermediate optimum, and it often comes as a surprise to dicover that perhaps a good deal of the scope for natural selection is taken up by selection in favour of keeping gene frequencies as they are rather than changing them.

The alternative forms of a gene which controls a particular function are called *alleles*; we refer to different alleles at a single locus or gene site. Directional genetic selection will change the frequency of such alternative alleles either at just one such site or at a number of them. Sometimes we can identify the genetic constitution of an individual in terms of particular alleles and, where this is so, we have the ideal situation for exact analysis and comparison between generations or between populations living in different localities, since we can express the differences quantitatively in terms of frequencies of alternative alleles. But more commonly we have to deal with situations where this is not possible, and we have only measurements of continuously variable traits, either in unrelated individuals or groups of relatives. In such cases we run almost inevitably into the problem of trying to decide how far genetic or environmental causes contribute to the differences. Also we do not know how many different kinds of gene are involved in such differences, nor the number of alternative alleles at the different loci. From the layman's point of view most of the interesting situations fall within the latter category. This has given rise to a great deal of confusion, since the definition of terms, the methods of analysis and the limits of valid inference require a degree of statistical precision which is unfamiliar to many who are interested in these questions.

Like any other species of animal, man has been shaped by the process of natural selection, but differs in one unique respect, namely the invention of culture, which intervenes between him and his habitat. Even in the most primitive contemporary societies, culture blunts the direct impact of the physical environment while, at the other end of the scale, people in sophisticated industrial societies live in an almost entirely man-made environment, while many gradations occur between the extremes. Given such a variety of life-style, generalizations are likely to be misleading.

Since, in the grand design of nature, the role of natural selection calls to mind the adaptation of animals and plants to their environment, perhaps the best way to approach the subject is to consider first the evidence for adaptation in contemporary man.

Human adaptation

The obvious physical differences which distinguish major ethnic groups are commonly regarded as primarily adaptive either now or formerly. The human species apparently agrees with other animals in following long-recognized ecological rules:

(i) The lower the average habitat temperature, the greater the body weight.
(ii) The higher the temperature, the larger the relative size of appendages like ears or limbs.

In comparative studies of human populations, Roberts has reported that about half the variance between populations in body weight and shape can be accounted for by differences in average temperature. We have only to recall the stocky compact build of Eskimos compared with the attenuated profile of equatorial Africans to be persuaded of the biological validity of these generalizations. Of course there are always exceptions in comparisons of this kind for which various explanations can be readily suggested, such as migration between climatic zones, with insufficient time for the migrants to have adapted, or cultural development may have removed the need for them to do so. The greater the cultural sophistication, the less the practical significance of formerly important physical differences.

Variation in degree of skin pigmentation is generally of adaptive significance in protection from ultra-violet radiation, the intensity of which is determined by the ozone distribution of the upper atmosphere. Adequate pigmentation protects against extreme exposure, which otherwise causes damage to the epidermis and affects heat regulation. Skin cancer of the exposed areas of the face and hands occurs much more commonly in unpigmented whites in tropical and semi-tropical countries than in the pigmented native inhabitants. Such differences are evident between whites and blacks in the southern United States, although the frequency of skin cancer in unexposed regions, e.g. under arm, is similar in the two groups. In zones of intermediate exposure, the situation is confused by secondary adaptive tanning, which can often produce phenotypic differences similar to those of purely genetic origin, thus providing a simple practical demonstration of the problems of distinguishing the effects of heredity and environment.

The application of reflectance spectrophotometry to measure small differences in skin pigmentation has demonstrated a high correlation between intensity of pigmentation and latitude. By taking account of this variable, as well as mean temperature and altitude, virtually all the variance in skin pigmentation between groups has been accounted for. However, there does not appear to be any report of differential survival or reproductive success related to pigmentation within populations, which is the hard evidence we should like to have for a demonstration of selection in action.

In passing, we might wonder why northern people should be so predominantly white. *Homo sapiens* is essentially a tropical animal, and no doubt the ancestral skin was pigmented. We do not know for sure why pigmentation was lost, but it may be related to the manufacture of vitamin D under the influence of ultra-violet radiation. In regions of weak sunshine, the presence of pigment would further reduce the manufacture of this essential vitamin, with consequent risk of rickets. This disease has been virtually eliminated in the white British population, but there have been a number of recent reports of increasing incidence of the consequences of vitamin D deficiency in children of immigrants with pigmented skin and, although there may be other causes, such as unsuitable diets, we cannot exclude vitamin D deficiency which is influenced by skin pigmentation in areas with low levels of sunlight.

Another physical feature which differs between ethnic groups is nose shape. This is believed to be adapted to the humidity of the inspired air, which influences the rate of flow of mucus secretion in the respiratory tract, and hence the clearance of bacteria, dust, etc. Thin noses which humidify the incoming air are found in dry, cold or hot environments. The broad noses of the more humid regions favour loss of water from the respiratory tract and promote cooling by evaporation. Some anthropologists regard the tightly curled hair of negroes and melanesians as adaptive, since it ensures a protective layer of air on top of the head. The so-called peppercorn hair of the Bushman is believed by some to be effective in promoting evaporative cooling. It has been suggested that the mongoloid type of face is adapted to intense cold; the bony ridges are reduced, with a corresponding reduction of frontal sinuses, nose prominence is diminished, and there is a redistribution of fat to produce better insulation, a syndrome of features most evident in Eskimos and North American Indians.

Physiological adaptation

Several attempts have been made to look for differences in tolerance to the extremes of heat and cold between ethnic groups regularly exposed to such conditions compared with others who are not. For example, Wyndham has carried out studies on resistance to heat on the part of the South African Bantu and northern acclimatized Caucasian whites under the climatic conditions of the South African veldt in winter. For experimental four-hour work periods, all the Bantu performed satisfactorily, whereas about half the Caucasians had to give up due to overheating or exertion coupled with an increase of deep body temperature, indicated by a rise of rectal temperatures of 40°C, unlike the Bantu whose deep body temperature was unaffected. But when the same whites were acclimatized by exposure to periods of very high temperature, then their performance resembled that of the Bantu in terms of work ability and deep body temperature, although the Bantu's sweat rate was lower than that of the whites. When similar groups of people were compared under cold conditions, there was no apparent difference between Bantus and whites, except that the deep body temperatures fell somewhat lower in the former; this was attributed to their lower skinfold thicknesses, which provided less insulation. Other studies of performance in very cold conditions underline the importance of acclimatization as well as general physical fitness. When comparisons are carried out between groups of people drawn from different climatic zones, there may be striking differences, but when such groups are allowed to acclimatize, then the differences may vanish or be greatly reduced.

The Australian aborigine presents an interesting example. Traditionally, the tribes living in central Australia were exposed to a wide range of daily temperature, which frequently fell to very low levels at night. Since the aborigines went around naked, and had only small fences and very inadequate fires at night camps, the question arises whether they are endowed with particular ability to withstand cold. There are two major kinds of possible adaptive change in response to cold. The reaction to comparatively minor reduction of the ambient temperature involves vasoconstriction, piloerection or gooseflesh, and shivering—which are adequate for moderate falls of temperature. With more drastic reduction, a critical level is reached at which metabolic changes take place, which lead to the release of heat and redistribution of blood to cooled areas—which, of course, may increase the rate of heat loss. In man, the critical temperature generally is about 27–28°C, comparable with what is found in tropical animals, but it has been noted that in the Australian aborigine the body temperature can drop to comparatively low levels, so the critical temperature may be lower. At comparatively low temperatures, the heat loss of the Australian aborigine may be less than that of whites. When aborigines were compared with whites at the other end of the temperature scale, there was no noticeable difference between them when both were acclimatized, except that the whites sweated more than the Australian natives, a difference reminiscent of that between Bantu and Europeans.

The Eskimos are famous for having evolved a sophisticated culture which enables them to live under the most inhospitable conditions. But there is also reason to suspect that they may be cold-adapted; we have already mentioned the mongoloid face with reduction of protruding parts which may contribute to their resistance to frostbite and ability to withstand low temperature—a property they share with North American Indians.

An interesting situation occurs in Indians living at 3000–4000 m in the Andes of Bolivia, Chile and Peru, where they are exposed to very low oxygen tensions and also to regular occurrence of low temperatures. While various cultural adjustments protect against the cold, it is not obvious how the effects of low oxygen tension can be mitigated by such means. Indians living at these high elevations display massive chest development with increased frequency of ventilation. They manifest a high work capacity compared with low-level Indians, and it has been suggested that larger capillary beds, greater myoglobin content in muscle, and higher anaerobic cellular respiration are contributory factors. It may be significant that these people have not encountered competition from other ethnic groups who may be less well adapted to these conditions and, while there does not appear to be any direct evidence in man, it is known that fertility is lower in various domestic animal species when transferred to high altitudes. On the face of it there would appear to be a clear indication of a syndrome of adaptive changes, but it is at present quite uncertain how far these are truly genetic rather than acquired changes. Studies are in progress which should clarify the situation.

A particularly interesting example of natural selection has probably been responsible for difference in tolerance to lactose among adults. This is the main carbohydrate in milk and is split by the enzyme lactase into glucose and galactose, which are readily absorbed. In infants, lactase is normally present in the small intestine; its rare absence is generally lethal and is due to homozygosity of a recessive gene. In older persons, lactase activity may vary according to whether they are members of a traditionally milk-drinking group. Although the domestication of milk-producing animals occurred in western Asia 5000–6000 years ago, as can be inferred from ancient drawings, ethnic groups vary greatly in their use of milk and, probably as a consequence of variable natural selection, in the activity of lactase in adults. The level of tolerance to lactase ranges from a few per cent in Europe to 80–90% or more in such groups as Australians, Chinese, Bantu, Thais and American Indians. It has been estimated that if the selection differential in favour of lactase tolerance were, say, 0.01, it would have taken 400 generations to attain the present high values in some populations. Since this exceeds the period of domestication, it would be necessary to increase the selection differential proportionately.

Another source of possible selection in contemporary man might be looked for in regional differences in diseases. Various infectious diseases and malnutrition have always been with us, but the hope that urbanization and improved conditions would eliminate diseases of all kinds has not been realized. We are all familiar with the statistics of coronary heart disease, lung cancer and bronchitis, and the increase in incidence of diabetes, as well as the increasing frequency of psychiatric disorders in wealthy western societies. But it is doubtful whether such diseases act as selective forces to alter gene frequency to any appreciable degree. In the vast majority of cases they become effective after the reproductive phase of life and therefore are irrelevant to gene frequency.

Natural selection is, of course, not the only factor which can influence gene frequency distributions in human populations. As a result of great variation in technological development, more technically advanced peoples have, over the course of history, thriven at the expense of less sophisticated groups who have been pushed into less fertile or climatically less favourable regions where they are exposed to greater rigours of natural selection. Such populations are often fragmented into small groups which may risk extinction in epidemics or natural disasters, with consequent loss of unique gene combinations. The colonization of North America by British and other European immigrants, followed by the decimation of the native Indians who were at such technical disadvantage, provides one of the most dramatic examples in recent times. It would be wrong to correlate technological differences between peoples with innate genetic differences in aptitude, since they are more properly regarded as historical accidents arising from the interaction between culture and habitat, as well as differences in degree of isolation. Such technological differences, often associated with mass migration, colonization or military adventure, have frequently resulted in grossly disproportionate proliferation of particular ethnic groups, with corresponding global changes in patterns of gene frequency, generally over much shorter periods than would ever be encountered within populations.

From this brief survey of major ethnic differences in morphological and physiological characteristics, a plausible case can be made for a major role of natural selection in establishing these contrasts which are most evident for people living under more extreme climatic conditions. The genetic status of physiological differences is relatively much more suspect than that of morphological or pigmentation differences. The human species is remarkably adaptable, and many of the physiological adaptations may represent functional plasticity during development, which leads to adaptive changes. If the stress continues for a sufficient number of generations, truly genetic changes may become established by the process of what Waddington called "genetic assimilation", whereby changes caused by selection take over, in part at least, the responsibility for controlling growth and development to produce the adaptive features, even in the absence of the primary environmental stimulus. So the situation may be very complex, with both genetic predisposition and non-genetic plasticity of response contributing to the characteristic array of adaptive reactions. If it is so often difficult to be sure how far striking differences in morphology or physiology are genetic, and also how far individual differences in such respect influence fertility and/or survival, we might anticipate formidable difficulties in measuring the incidence of natural selection in urban societies in which the range of environmental conditions is vastly less and the complexity of the protective culture is correspondingly greater.

Secular change in morphology

In various parts of the world, burial customs have provided anthropologists with large samples of skeletal remains extending back for several centuries. Radiocarbon dating makes it possible to compare their shapes and sizes according to their age. Most information is available for cranium shape, which refers to the ratio of maximum width to the maximum anterior/posterior dimension. From several parts of the world there is evidence of a trend to increasing broad-headedness or brachycephaly, which seems to have been going on since the end of the Pleistocene. Such progressive trends pose the usual problem of genetic or environmental origin.

To take a particular case, in Eskimo and Aleutian populations there has been a substantial increase in head width. In early settlements in west Alaska, the index of head width was 73 compared with the contemporary average score of 83. Internal evidence, especially from a study of dentition, suggests that the changes took place within the population and that immigration is unlikely to have been the cause. Radiocarbon dating suggests that it took place progressively over a period of about 1500 years, i.e. about 50 generations.

In other studies in east central Europe, the skeletal material goes back about 700 years during which time there has been a similar change of about 10 units in favour of brachycephaly. Bielicki and Welon have looked for direct evidence of natural selection by comparing the numbers of living brothers and sisters of recruits to the Polish Army in the early twentieth century when cranial and other dimensions of recruits were recorded. They found that individuals of intermediate head width had the highest number of living brothers and sisters, compared with the most extreme narrow or broad-headed persons, so there appeared to be evidence for stabilizing selection which would tend to keep the head shape constant. But, in addition, the curve relating number of living sibs and head shape was found to be skewed in favour of a higher number of living sibs in the most broad-headed individuals, so there was also evidence of natural selection in favour of the broad heads. For the population concerned, ethnic admixture was considered unlikely. As the authors reasonably suggest, although the evidence does not prove it, the occurrence of selection seems likely. At the time when the recruits were born, living conditions were harsh, infant mortality was very high, and life expectation comparatively low. With the improvement of living standards and the drastic lowering of infant mortality, it is possible that former natural selection which influenced head shape has become very weak or perhaps ceased to act.

The stress imposed by unfavourable living conditions is an important consideration. It is possible to estimate age of death fairly accurately from skeletal remains, particularly by the state of the epiphyses and the degree of closure of the cranial sutures. In earlier times only a few per cent of the population exceeded the age of 40 at death. Angel, in a long series of Greek skulls, has shown there was a very slow increase in longevity during the period of early civilization. Karl Pearson estimated the expectation of life in ancient Egypt at about 22 years from the ages indicated on mummy cases. Where life expectation is very low, there is far more scope for natural selection than in contemporary developed societies.

In recent times there is plenty of evidence of changes in stature as a consequence of immigration to a country with higher living standards. This has often been demonstrated in comparisons between parent and progeny of immigrants to the United States from deprived parts of the world. A particularly clear example of this was described by Shapiro who compared the size and growth rate of the children of Japanese immigrants to Hawaii and found clear differences, no doubt due to better conditions during growth. But he gave the analysis an unexpected twist by recording the size and body proportions of relatives in Japan, and compared them with data from families from the same villages who had not produced emigrant individuals and found a similar trend, which suggests that migrant individuals may not be a random sample of the population. If such differences are genetic, large-scale emigration from a population may amount to a form of selection.

Damon and Thomas, in a study of height and body build in 2616 Harvard college men who had completed their reproductive life, reported that those who remained single were 0.8 cm shorter than those who married and had children, while those who married and had no children were 1.0 cm shorter than the others who did. A recent re-analysis of the data suggested that the evidence is more compatible with stabilizing selection favouring intermediates rather than a directional trend, although the evidence is inconclusive.

Perhaps the most convincing demonstration of apparent stabilizing selection refers to the relationship between birth weight and infant survival. Karn and Penrose, and van Valen in large-scale studies on survival of new-born babies in London and New York have shown that, as birth weight deviates, in either direction, from the average value the survival rate declines—exactly the relationship between fitness and deviation from the mean expected in stabilizing selection. However, it is uncertain how far the genetic reality corresponds to the phenotypic appearance, since a baby's birth weight is mostly determined by maternal influences which are non-genetic. But even if a small fraction of the variability in birth weight is heritable, this will tend to reduce drift from the mean. It is also possible that the effectiveness of such a stabilizing influence is conditional on environmental conditions. In favourable environments babies with higher or lower birth weights, who would not have lived had conditions been more adverse, may now do so and, if there is a genetic component in such differences, this would tend to favour an increase in the genetic variability of birth weight.

So far we have dealt with characters which are continuously distributed, whose inheritance is regarded as polygenic, i.e. controlled by many different genes, and whose variation can only be handled by biometrical methods. Where we recognize the existence of genetic differences, we infer differences in gene frequency, but generally have no idea of the number of genes involved nor the distribution of allele differences at the loci concerned. The technique of analysis is based on the measurement of resemblances between relatives and, if the situation appears compatible with the model upon which such an analysis is based, we can estimate the relative importance of heredity and environment in the variation between individuals of the population. However, there is one important qualification in the application of these methods to man compared with livestock or laboratory animals, namely that we can never test their accuracy by carrying out artificial selection to see whether the response to selection agrees with the estimates of heritability, as we can in livestock and laboratory animals. Therefore there is always a degree of assumption in extrapolating from observed correlations between relatives to the situation in the general population.

If we are dealing with differences in, say, growth rate or serum cholesterol concentration or some other trait which can be directly equated with the corresponding character in laboratory animals or livestock, we can often take the estimates of heritability provided by the latter as a fair guide as to what we might expect in man. Enough selection experiments and progeny tests have been carried out to demonstrate that heritability is generally fairly high in characters which vary about an intermediate optimum, and tend to be low or very low in fitness traits like fertility. The real difficulty arises when we transfer our attention to measures of behaviour, like intelligence quotient, which have no valid counterpart in other species and for which the act of extrapolation from correlations between relatives involves a greater act of faith.

Single gene differences

A rather different analytical situation arises when we turn to the examination of single gene differences identified by antigenic or electrophoretic methods, since we can compare populations in terms of allele frequencies at particular loci. All such variation arises in the first instance as an error in DNA replication, leading in the vast majority of cases to insertion of the wrong base at a particular point in the base sequence of a given gene and hence replacement of the usual amino acid by a different one, unless the new base sequence codes for the same amino acid due to degeneracy of the code. The great majority of such mutations are lost by chance quite soon, unless they confer a very substantial advantage—in which case they may spread through the population to reach fixation. By virtue of the Mendelian type of inheritance and the diploid/haploid alternation, chance plays an important role wherever genetic variation is present. The smaller the population size, the more important the chance variation in gene frequency. It is not particularly difficult to visualize how new alleles which cause no damage in fitness may reach a high frequency in the course of time by the hazards of sampling alone.

The corner stone of population studies of gene frequency is the Hardy-Weinberg law which states that the gene frequency remains constant provided mating with respect to genotype is random, there are no differences in fitness, and that the population size is large enough to make sampling variation unimportant. If we can identify the genotypes in a two allele situation (a, b) so that p, q represent the gene frequencies such that $p+q = 1$, then the distribution of respectively one kind of homozygote (aa), the heterozygotes (ab) and the other kind of homozygote (bb) will be represented by

$$(p+q)^2 = p^2 + 2pq + q^2.$$

Many studies of gene frequency in different species have validated the generality of this relationship.

In the identification of gene differences by electrophoresis, an electric current is passed through a sample of serum or an extract from some tissue which is applied to a suitable substrate (such as a layer of starch, acrylamide or agar gel) so that, as the current flows through the system, molecules in the extract which differ in charge move at different speeds away from the point of origin. After the process has been allowed to go on for long enough to achieve a convenient separation of the molecular species in which we are interested, then, if the proteins are present at a sufficient concentration, staining for protein will indicate a series of bands at the positions to which the molecules have migrated. However, in most cases we are interested in enzymes which occur at too low a concentration to be identified in this way, and so we adopt a different procedure. After the molecules have been separated, they are allowed to react with a substrate specific for the enzyme concerned, and the product of the reaction then interacts with a suitable stain so that the result of the reaction between stain and product is laid down as an insoluble complex in the gel, in the zone to which the enzyme molecule has migrated. In this way we can identify the enzyme as a coloured band or bands at characteristic distances from the origin. The application of this procedure has revealed that many proteins, including enzymes, occur in two or more forms, which migrate to different distances in the gel and which are determined by different alleles. The alternative alleles occur with quite high frequency in particular populations, and such populations are said to be *polymorphic* for the locus or gene site in question. Comparisons between populations have often demonstrated differences in frequency of the alternative alleles. In many of the cases so far investigated, these polymorphic differences arise from a single base substitution leading to a single amino-acid substitution in the polypeptide concerned.

The occurrence of such population differences, and of polymorphism in general, involve the usual problem as to whether differences arose by either natural selection or by hazards of statistical sampling, or both. The allelic changes arise by mutation. Over evolutionary periods, random sampling of the genes transmitted from generation to generation may have led to unpredictable fluctuations in frequency of alleles which do not differ in their effect on fitness. On the other hand, that may not be the explanation, and the alleles may affect survival and/or reproductive performance in some way and be thereby subject to selection. This is a large and controversial subject which we cannot follow here. We shall confine our attention to instances where there is a strong indication of a selective origin of population differences in simply inherited alternatives.

The most convincing examples of the influence of natural selection on gene frequency are encountered in resistance to malaria. Although different genes are involved, causing different kinds of change in the red blood cells which are attacked by the protozoan *Plasmodium* parasite, the geographical distribution is often closely similar and they share certain genetic characteristics which probably make them valid general examples of how disease resistance arises and is maintained.

The principle forms of resistance are as follows:

Sickle-cell anaemia

This is the classic case so often described, in which there is an electrophoretic difference in the beta chain of haemoglobin caused by a single amino-acid substitution of valine in place of glutamine in the amino-acid sequence, and this in turn depends on a single base substitution in the DNA. The amino-acid substitution has far-reaching consequences, since the modified haemoglobin is less soluble in the deoxygenated state and tends to precipitate under such conditions to form crystals which distort the shape of the red cells—hence the name *sickle-cell haemoglobin*. Persons who are homozygous, i.e. have two doses of the sickle-cell allele, characteristic of many African populations, suffer from acute anaemia which is generally lethal before reproductive age. Where this is apparently not so, as in some other tropical regions, a different allele is probably present.

Allison, who has done most to elucidate the situation, has shown that in children heterozygous for the sickling trait the frequency of infestation by the malaria parasite is less than in children who are homozygous for the normal allele, while women heterozygous for the sickle-cell allele may be more fertile. Misunderstanding has occurred in the past, since differences in survival are generally not evident in adults due to the occurrence of acquired resistance to malaria, another example of the confusing effects of non-genetic acclimatization. Hence the high frequency of the sickling allele where malaria is a serious risk is due to the superior survival of heterozygotes, which is sufficient to offset the genetic consequences of lethality of the homozygous condition. The polymorphism is maintained by a balance of selection pressures which work in opposite directions. If malaria is eliminated by extermination of the mosquito carrier, or a population or sample of it migrates to a new district free of malaria, then the sickle-cell condition becomes an unqualified genetic disease, as in Americans of African origin in whom various steps are being taken to mitigate the consequences of segregation of the sickle-cell allele. This gene reaches its highest frequency in regions of West and East Africa, although it is widely distributed in central regions and also turns up in regions of India and further east.

The thalassemias

The array of genetically determined forms of anaemia, known collectively as thalassemia, is of more complex origin than sickle-cell anaemia. In this case there is lack or deficiency in the synthesis of red cells which have lower osmotic fragility. Deficiency of the alpha chain of haemoglobin in homozygotes of α-Thal-1 is due to partial or complete absence of the gene for the α-chain. As a result of mutation it has been deleted from the genome. Other, rarer forms of α-chain deficiency may be due to base changes in the controlling gene, which may affect the stability of the messenger RNA which carries the coded information for alpha-chain synthesis. There is also another form (α-Thal-2) which may, like α-Thal-1, involve lack of part of the gene for chain synthesis.

Deficiency or lack of beta-chain formation is likewise complex in origin and may arise from similar causes, i. e. deletion of part or all of the gene concerned, as well as other changes which may affect chain synthesis. The situation in thalassemia is thus complex but is likely to be cleared up in the near future.

The anaemic condition slows growth and causes facial differences due to bone thickening which accompanies increased bone marrow activity. This is widespread in parts of the world exposed to malaria, especially parts of the Mediterranean, Thailand and New Guinea. In Cyprus and Turkey 15–17% of the population is heterozygous for beta-chain deficiency (Cooley's anaemia), hence infant mortality due to homozygosity is high. It is generally accepted that such high frequencies are due to selection of heterozygotes in areas prone to malaria. It has been estimated that a heterozygote for the beta chain has approximately 9% greater chance of survival to the reproductive stage of life in the affected regions. Thus, this genetic condition presents a major problem of public health in the areas concerned in which the difficulties of identifying the origin of suspected genetic anaemia are increased by the occurrence of nutritional anaemia, as well as other genetically determined differences in haemoglobin.

G6PD deficiency
This is due to a recessive gene which leads to loss of activity of the enzyme glucose-6-phosphate-dehydrogenase. This gene is sex-linked, which means that there are two types of male, normal and affected, while in females there are three alternative genotypes and an overlapping distribution of effects. In affected individuals there is a deficiency of reduced glutathione which is required for stabilization of the red-cell membrane. Recognition of this genetic effect was aided by the discovery that when certain persons, who turned out to be G6PD-deficient, were administered various anti-malarial drugs, this led to rapid destruction of red blood cells. During the Korean War, for example, it was noted that such haemolysis frequently occurred in US blacks treated with the anti-malarial primaquine, but occurred only rarely in whites. Also it has long been known in the Mediterranean, where G6PD deficiency is fairly common, that eating beans (*Vicia faba*) has serious consequences, known as favism.

Distribution of the G6PD deficiency closely parallels the regions of high malarial risk. In many parts of tropical Africa, two genetic varieties occur at high frequencies with either slight or mildly reduced activity, due apparently to differences in rate of decay of the enzyme in the red cells. Hence in young cells there is no difference between the alternative genotypes. A great number of different forms of G6PD have been reported, probably indicating a series of alternative alleles, and the interpretation of this variation is still very much a research problem.

In Greece, Cyprus and Sardinia, the correlation between occurrence of G6PD deficiency and risk of malaria has been shown to be particularly high, especially in Sardinia where the coastal regions, normally prone to malaria, have the highest frequency of the gene compared with inland areas at high elevation which are both comparatively free of malaria and the gene for G6PD deficiency.

Thus the three major groups of polymorphisms (sickle-cell anaemia, the thalassemias and G6PD deficiency) represent alternative solutions to the problem of how to achieve resistance to malaria. They have in common the provision of an unfavourable environment for the malarial parasite in the red cells. In sickle-cell anaemia it has been suggested that the parasite is adversely affected by the presence of the alternative haemoglobin, and also that the existence of parasites in the cells may cause them to stick to capillary walls long enough to use oxygen—which promotes sickling, followed by greater risk of phagocytosis. In G6PD deficiency, growth of the parasite is interfered with by reduction in the availability of reduced glutathione, particularly marked in older cells which are preferentially parasitized.

The polymorphisms also agree in requiring the assumption of superior fitness of heterozygotes in malarial regions, so they are balanced polymorphisms, although only for sickle-cell anaemia has the superior fitness of heterozygotes been directly demonstrated.

Blood groups

Of the many independent blood-group systems now known, the ABO group was the first studied, and determinations of these blood groups have now been carried out in all parts of the world on a total of something like seven million people. Table 3.1 indicates the levels of polymorphism in some major ethnic groups. There are very striking differences in gene frequency which, together with other blood-group differences between populations, have been used to work out degrees of genetic similarity and reconstruct possible former migrations. Thus Beckman, in a study of gene frequency in Sweden, has claimed that most of the regional variation can be explained by contributions from three ancestral strains, an east European, a west European and a Lapp strain, which differed in gene

Table 3.1 Percentage frequency of ABO blood group alleles in different ethnic groups (based on figures from *Man's Natural History* by J. S. Weiner, Weidenfeld and Nicolson)

	Gene Frequencies		
	A	B	O
South African Bantu	19	12	69
Hottentot	21	16	62
Carib Indian	4	1	95
Eskimo	25	2	73
Chinese	20	24	56
Australian aborigine	37	0	63
Micronesian	20	10	70
Brahmin	25	20	55
Afghan	21	25	54
English	25	5	70
Bedouin	17	0	73

frequency and which tended to occupy different areas of Sweden. Regional differences in frequency can be attributed to colonization by small groups with an atypical gene frequency who have been subsequently reproductively isolated from the parental population and have multiplied to establish a strikingly different array of alleles.

In studies of this kind it is inferred that the ABO blood groups are practically neutral in their effects, and merely reflect statistical sampling and past migration. However, there are also strong indications of differences in the frequency with which alternative blood groups are associated with particular diseases within populations. Thus Clarke has described a very definite association between duodenal ulcer and also stomach ulcer and type O; cancer of the stomach is associated with type A, and likewise pernicious anaemia. This evidence does not help us to account for the maintenance of the polymorphisms, however, since the possible selective effects would be very weak as these diseases generally act after the reproductive period of life. But we cannot ignore the possibility that the ABO blood groups may have played a former role in resistance to infectious disease, perhaps with superior fitness of heterozygotes when the population was exposed to infection. Certainly the ABO system appears to be of long standing, since it also turns up in a number of primates.

Serological differences

A further source of variation is provided by the growing number of immunological differences detected by serological methods which have revealed arrays of differences between populations and pose similar problems as to their origin. Of particular interest is the variation at the closely linked histocompatibility loci which control differences in degree of graft rejection. A variety of antigen whose presence can be ascertained by special tests which make use of lymphocytes, is determined by alternative alleles at several closely-linked loci. In fact, they are so close together that within any population certain pairs of alleles are inherited together. Since breakage and exchange of chromosome regions which carry the linear sequence of genes (or *crossing-over* as it is called) is a regular feature of the cell divisions preceding gamete formation, we expect, in due course, independence in the distribution of the alleles and hence antigens for the two loci. In fact, we find the presence of statistically improbable associations, which suggest that particular allele combinations may have a selective advantage over other combinations.

The differences between populations in the allele frequencies have also aroused interest and generated the familiar speculative discussions. It is particularly suggestive that certain diseases like Hodgkin's disease and ankylosing spondylitis are associated with the presence of particular alleles. This is an active area of research, and it is anticipated that further associations of this kind will be detected.

In addition to the histocompatibility loci, a long series of immuno-logical differences in the serum globulins have been recognized, and this information is being used for a more detailed genetic description of populations and ethnic groups which will not only shed light on their genetic affinities but at the same time provide better criteria for deter-mining whether selection has played a role in different polymorphic systems.

The general conclusion from these studies is that we have now some well-established examples of differences in gene frequency produced by natural selection, especially via resistance to malaria. In many other polymorphisms the evidence for selection is only suggestive or entirely non-existent. Since infection by different pathogens probably constituted a large part of what we regard as the unfavourable environment of former times, elimination of such diseases has naturally meant that selection for resistance has ceased. We may therefore be left with either unqualified genetic diseases manifested in homozygotes, or virtual lack of difference between the effects on fitness of the alternative homozygotes and the heterozygotes, in which case the inertia of the Hardy-Weinberg distribution and the hazards of sampling will chiefly determine the subsequent course of gene frequencies.

Induced mutation

Concern is often expressed at the risk of increase in mutation from increased exposure to radiation or mutagenic chemicals as a result of

Table 3.2 Natural and artificial exposure to radiation in biological units of dosage (based on values quoted by I. M. Lerner and W. S. Libby in *Heredity, Evolution and Society,* 2nd edition, W. H. Freeman & Co., San Francisco, 1976)

Radiation Source	Average dose
Natural	
Cosmic radiation	30–60
Earth radiation	50–60
Natural radiation (ingested)	20–30
Total natural	100–150
Artificial	
Diagnostic, medical and dental	30–75
Occupational	0.2–0.8
Miscellaneous	<4
Total artificial	32–80

environmental pollution or as a by-product of industrial or medical practices. A great deal of effort has been devoted to measuring exposure to radiation and determining maximum permitted levels of safety, based on measurements of mutation rate, and their effects in fruit flies, mice, microorganisms, etc. Table 3.2 summarizes the relative importance of normal unavoidable levels of radiation and the additional exposures arising from human activities. Such estimates, which refer chiefly to urban western societies, are subject to a very wide range of variation. The values are expressed in terms of units of a scale which relates biological effect to a degree of exposure measured in physical terms. Natural radiation which includes cosmic radiation, earth radiation (such as that emitted from different types of stone used in building) and ingested natural radiation, accounts for two to three times the artificial exposure, of which the diagnostic use of X-rays in medicine and dentistry comprises by far the major part and underlines the need to minimize the exposure of patients and hospital staff to such risks of increased mutation. Children who spend a great deal of time viewing television at close quarters constitute another group at risk. It has been estimated that they may be expected to receive radiation equivalent to about one third of the natural dose per year.

The risk of increased mutation is not confined to radiation, but also embraces chemical compounds which can interfere with the process of DNA replication and cell division, and thus may cause base changes or chromosome breaks and alterations which, if they occur in the germ cell, may lead to sterility or abnormal growth in the progeny. Apart from such strictly genetic effects, these compounds are generally carcinogenic as well. Drugs, pesticides, herbicides and other chemicals used in agriculture are particularly likely to act as mutagens, and methods are being devised in many countries at present to monitor the possible mutagenic or carcinogenic effects of new compounds which are introduced into industry. But there are many parts of the world where indiscriminate use of dangerous chemicals, often by spraying from the air (with consequent contamination of the skin and water supplies), may be generating an unmeasured hazard.

All human activities which increase mutation rate in different ways call for regular monitoring. This is comparatively easy for radiation in which small doses can be measured accurately; it is more difficult for the drugs and industrial chemicals which are produced in such vast numbers and which may create risks after they have been broken down in the body to other compounds.

Selection for the future

Having summarized the kind of practical evidence which has been presented in favour of natural selection, and appreciated the difficulties in demonstrating its occurrence directly, we can now turn to the more general topic of the likely consequences of selection pressure in contemporary and future society. Such observations are inevitably speculative and lean heavily on the theoretical deductions of population genetics, although the predictions of social commentators often dispense with such discipline. It is very sensible for any thinking person (conscious of the part natural selection has played in making man what he is) to wonder whether the process is still going on and, if so, whether we can discern any changes in the way it works and what meaning this might have for the future.

The first point to recognize is the substantial reduction in the scope for natural selection if we compare the situation in present-day urban societies and past times. Effective selection depends upon variability in fertility and/or survival. If everyone survived to post-reproductive age, and if everyone married and produced two children, there would be no selection. The gene frequency might drift around by chance, but there would be no systematic alteration. No population meets these stringent requirements, but the more affluent western societies have approached nearer to them than probably any other population in the history of man.

Demographic comparisons in the United States show how dramatically the opportunity for natural selection has declined in the last century. Thus in 1840 only 62.8% and 66.4% of males and females respectively reached the age of 15, while in 1960 the corresponding figures were 96.8 and 97.5%. In 1840, 56.2% and 58.1% of males and females reached age 30; by 1960 the figures had risen respectively to 95.1 and 96.9%. As these two ages span a substantial part of reproductive life, the scope for natural selection by differential mortality has been substantially reduced.

Dudley Kirk has discussed the demographic changes which affect the scope for selection via fertility under the headings of mating and marriage, childlessness, number of offspring, age of child bearing and average generation time. In the United States the married state has become more common in recent years due to the greater proportion of people who marry (and also the earlier age of marriage) so that a greater part of the reproductive life is spent in the married state. Thus in 1910, 88.6% of women married by ages 35 to 44 compared with 91.4% in 1940. The median age of marriage fell from 26.1 in 1890 to 22.8 in 1960.

Childlessness may be due to either low fertility, failure to mate, or birth control. The combined effect of mortality and childlessness meant that about half the women born in 1840 did not contribute to the next generation, compared with the present 15% or less.

The average number of children per family is now much lower than formerly—two to three compared with the earlier average of seven. The decline in average family size does not by itself reduce the opportunity for natural selection, which is determined by the variability of family size, not by the mean. It turns out that the scope for selection was actually higher for women born in 1900 than those born in the 1870s, so the scope for natural selection may be uncorrelated in different components of fitness.

The pattern of reproduction has also changed. In the United States it has been recorded that the median American woman has her first child at the age of 21 and her last at 27. Such a change in the age of reproduction has an influence on the appearance of those genetic diseases which are correlated with maternal age or number of children in the family and, since such effects are more commonly encountered in children of older mothers, this change has been beneficial.

These dramatic changes have been spreading throughout the social classes, and between urban and rural groups which formerly differed. Thus in 1900 white married women with less than eight years of schooling had more than three times as many children as those who had spent four or more years at college. By 1970 it was reported that the comparable difference was now less than 40% and the steady evening out of differences continues. In the US black population, the situation is more like that of the whites of former days, so that the opportunity for natural selection remains correspondingly greater—but the same trends are evident and all

Table 3.3 Opportunity for selection in three Chilean populations
(based on figures from J. F. Crow in "The quality of people: human evolutionary changes,"
pp. 309–320 in *Natural Selection in Human Populations*, ed. C. J. Bajema, John Wiley & Son
Inc., N. Y. 1971)

	Town	Village	Nomad
Average number of children	4.3	5.9	6.1
Variance in number of children	8.5	7.5	6.4
Proportion surviving to adulthood	0.87	0.75	0.42
Indices of opportunity for selection:			
Mortality	0.15	0.33	1.38
Fertility	0.45	0.22	0.17

the indications are that in urban western industrialized societies, the opportunity for natural selection has been or shortly will be reduced to levels never hitherto encountered by the human species on such a large scale.

Given the low survival rate in former times and the high infant mortality due to infectious diseases and other unfavourable conditions, inborn resistance would have been selected for, just as resistance to malaria is favoured in present-day communities at risk. There is no reason to doubt that similar polymorphisms and alternative genetic solutions would have arisen, and we may wonder how far present-day variation in blood groups, immunological variants and other polymorphisms represent the aftermath of such past selection.

The scope for selection is directly related to variation in fitness, which is determined by rates of survival and fertility. Crow has introduced a useful index of opportunity for selection which is the ratio of the variance to the square of the mean number of offspring. The global index can be subdivided into indices of selection determined by variation of mortality or fertility, and this allows us to compare populations either in time or space in terms of the potential scope for selection. For example, Table 3.3 shows the differences between social categories in Chile and highlights the differences in scope for selection by either fertility or mortality in the urban compared with the rural or nomadic groups. Similar comparisons over time in American white and black populations have shown a progressive decline in the level of the index which is more marked in whites than in blacks, in whom there is greater variation in both fertility and mortality. Whether such differences have an effect on gene frequencies depends on the heritable component of the variation and whether the selection is directional or stabilizing. We have so little information about such questions that we can hardly attempt rational answers apart from suspecting that the heritable component is likely to be secondary, perhaps minor.

Table 3.4 Survival and fertility rates in different countries. All figures are expressed per 1000 of the population. Infant mortality refers to death within the first year of life.
(based on figures from *The Genetics of Human Populations* by L. L. Cavalli-Sforza and W. F. Bodmer, xvi + 965 pp., 1971, W. H. Freeman & Co., San Francisco)

	Live births	Infant mortality	Death rate
Nigeria	50.1	76.2	12.8
Mexico	45.9	74.4	12.5
Jordan	40.1	63.1	8.1
Ceylon	36.6	57.5	20.3
Chile	35.5	127.7	12.5
USA	24.6	26.4	9.4
Australia	22.6	21.5	8.8
France	18.4	25.2	11.8
Japan	18.2	33.7	7.8
England and Wales	15.9	22.2	11.6
South Africa			
Caucasian	29.7	28.7	8.6
African	130.1	120.6	16.2

Table 3.4 gives an idea of the variation in live birth rates, infant mortality rates and death rates between different countries, and indicates the gross difference in scope for selection between developed and under-developed countries. Since life expectation has been so greatly increased, variation in fertility has become the major instrument of potential selective change in the most developed countries.

Evaluating the role of natural selection in human populations, and estimating the genetic contribution to various forms of ill health is bound to be a slow and laborious process fraught with many pitfalls, chief among them being the spurious correlation. More information is available for the human species than for any other, but most of it is scattered through unrelated files accumulated by different Government departments. What is needed is a system of computerized cross-linking of files to give access to comprehensive vital statistics of family groups over extended periods. An impressive start in this direction has already been made in parts of Canada, notably British Columbia. Such a development is unavoidable if we are to get a better understanding of familial proneness to diseases and how far the pattern is influenced by social class, occupation or ethnic origin.

Extrapolation from incomplete data often motivates political action or propaganda. Family planning was first practised in the better-educated wealthier sections of society, and gradually spread down the economic grades. When there were considerable differences in family size between "upper" and "lower" classes, eugenists expressed concern about the dysgenic selection which they thought inevitable. Since then, fertility differences have dwindled and the concern has subsided, accompanied perhaps by a greater caution in making *a priori* judgments about presumed genetic differences. Nevertheless, in spite of the present trend to reduced variability in family size, there remains sufficient variation in this respect in present-day urban societies to provide a basis for potential selection.

Such discussions have been given a new dimension by recognizing assortative mating as a factor of increasing importance. With the general availability of education, and the sorting out of people by the educational process into grades of ability, there will be a tendency for husbands and wives to be drawn from the same educational class. There is no doubt about the reality of assortative mating in, say, stature and it is a matter of common observation, supported by statistical measurement, that college students tend to find their marriage partners among college students rather than those who left school as soon as they could, and vice versa. One of the genetic consequences of assortative mating is to increase the level of genetic variation beyond what would be expected if mating were random; there are more homozygotes and fewer heterozygotes than expected. Such an enhancement of the genetic contribution to variation would lead to progressive reinforcement of the process if scholastic attainment has a genetic component, as many believe to be true of the intelligence quotient (I.Q.). Our educational procedures and social customs would be creating intrapopulation differences unique to our society, and could conceivably lay the basis for a divisive caste system. Given assortative mating, the more successful our educators in sorting and classifying types of ability, the greater the spread of genetically determined ability in the population.

That is one scenario but there are so many "ifs" and "buts" in it that undue concern is hardly necessary. Thus we remain rather ignorant of how far scholastic ability is genetic. Assortative mating *per se* does not involve selection, merely a difference in how the genes are distributed in the population; the variability would increase, but the mean would remain the same. If economic rewards were correlated with performance in our educational system and thereby determined larger family size, then we should have a potential pressure of selection. But reproductive fashions change and economic rewards and/or fertility may be poorly related, if at all, to the schoolmaster's grading scheme so that an apparent tendency to directional selection may be short-lived and ineffectual.

In recent times perhaps the most likely place to look for the action of natural selection is in the movement of peoples from a rural environment to large industrialized cities. Exposure to different kinds of disease, psychological stresses, requirements of different behaviour patterns, changes in diet, all add up to differences which, if we were dealing with animals, we should recognize as belonging to different habitats or niches, and would expect to find corresponding adaptations. But man is an adaptable species with cultural flexibility and, although we may not wish to dismiss the possibility, we have no satisfactory way of knowing at present whether such selective changes are taking place.

Although it might seem that primitive societies offer promising material for the study of natural selection, Neel (from his experience of studies on Amazonian Indians) has underlined the practical difficulties, such as the problem of obtaining large enough samples to arrive at statistically valid generalizations, the obstacles in the way of biomedical study in the field, and the resistance of highly trained experts to live for many years in the bush. Since the way of life of these remote primitive communities, which have arrived at a reasonable equilibrium with their environment, is on the point of destruction, we are likely to lose forever information which could help us to understand the changing pattern of selection in relation to social habits and community organization.

Perhaps the commonest recurring topic in the discussion of the genetic fate of man is concerned with whether the improvement of environment and hygiene—many individuals now survive who would formerly have died—is leading to a deterioration of the race or, put more precisely, potentially disadvantageous alteration of the average genotype. Medawar has dealt rather effectively with this topic by noting that it has two aspects. On the one hand there is the progressive amelioration of conditions and the alteration of the environment which may require a different genotype for an equivalent degree of adaptation and, on the other, there is the consequence of preserving genotypes which formerly would have been eliminated. With respect to the former, it may be argued that a good deal of what we call adaptation to a former adverse environment depended on monogenic metabolic changes which conferred resistance to specific disease organisms which have now been eliminated. Again we can make use of what we have learned from present-day resistance to malaria as a practical guide. Medawar makes the point that such resistance generally depends on "cheap genetic tricks", i.e. modifications of one or two metabolic stages which happen to interfere with the life cycle of some specific disease organism, but which are otherwise irrelevant to the growth, development or behaviour of the host. They are pragmatic solutions which have been favoured because certain mutations conferred higher fitness on their carriers in certain conditions, and this had statistical consequences for gene and genotype frequency in the population. Natural selection does not display foresight; an ingenious biochemist could probably devise better solutions which would not only preserve the carrier from attack of the pathogen but would also enable the homozygotes to survive as well. The Mendelian type of inheritance puts a premium on changes which produce their effect in heterozygotes, hence the "cheapness" of the genetic trick.

Apart from disease resistance, the environment may be changed in other respects, so that we may have lost or be losing adaptation to former conditions. But if they are unlikely to recur, the loss of adaptation is irrelevant. In this context we should not overlook the considerable inertia of gene frequencies.

The second consideration turns on the preservation of homozygotes of rare recessive genes like phenylketonuria, galactosemia, fibrocystic disease of the pancreas, etc., in which modern medical care can allow individuals of particular genotypes to survive and reproduce. This means that the gene frequency in such cases will, in time, become higher than would otherwise be so. Hitherto, natural selection, by discriminating against the homozygote, has kept the gene in check against the pressure of mutation; if the control is removed, the frequency must increase. Crow has computed the magnitude of the threat. If we are dealing with, say, a recessive gene with a frequency of 0.01, so that 1 in 10 000 of the population shows the genetic disease as a homozygote, then it can be shown by simple algebra that if such individuals are given a normal chance of survival and reproduction by medical care, the proportional increase in gene frequency will be 2% per generation, so that it would take about 35 generations, i.e. more than 1000 years, to double the gene frequency and raise the frequency of the homozygotes from 1 to 4 in 10 000. With such a time-scale compared with the rate of increase in scientific and medical understanding (which may find ways of dealing with such genetic effects by intervention during development or at the primary level), undue concern about the dysgenic effects of medical advance seems hardly necessary. But, ultimately, the consequences of increased frequency of harmful genes will have to be faced.

The application of genetic counselling, whereby carriers of such recessives are advised against marrying another carrier, would of course reduce the frequency of the homozygous types who manifest such recessive diseases but would not reduce the frequency of the gene. So as far as the population is concerned, the situation is not improved. If heterozygotes could be identified in adults and in foetal cells, a screening system for parents, amniocentesis in pregnancies at risk, followed by abortion of heterozygotes or homozygotes, would reduce the gene frequency in the population. But such a procedure could not be justified on ethical grounds, quite apart from the practical difficulty and expense of carrying it out. Of course the situation is rather different when we are dealing with dominant effects or chromosomal abnormalities which produce severe mental retardation, e.g. Down's syndrome. It is generally conceded that identification of an affected foetus, followed by abortion, is the humane way to deal with such genetic errors.

Finally, there is another, generally overlooked source of variation, namely the number of fertilized ova which die. James has estimated that about 49% of fertilized eggs do not survive until term. The greater part of these are lost within the first month after conception. The causes of this wastage are not fully understood; no doubt mistakes in development account for a proportion of early deaths. Strong selection for adaptation to the uterine environment might be expected, but it remains an interesting speculation how far changes in life style, diet, presence of drugs in the body fluids, etc., might influence the chances of implantation and survival according to the genotype of the zygote, and whether such selection might play a hidden role in altering processes whose effects are not confined to uterine life.

Thus to the question: is selection going on and, if it is, what is it doing to the human species? we cannot give a confident answer, even in genetically simple situations. It has been difficult enough to produce scientific proof for genetic resistance to malaria. Argument by analogy, and appeal to the correlation between gene frequencies and the geographical incidence of the disease, have played a large part in the story. Perhaps the most important message is that consistent changes in gene frequency are likely to be very slow. The environment can change out of all recognition without any obvious changes of genotype. Differences in reproductive habits between classes, religious groups or other intra-population categories, may lead to differences in the rate at which their genes are propagated relative to other groups, but persistence of such differences is often confined to a few generations which, though long enough in human terms, is trivial on the evolutionary scale. The effects of selection on gene frequency may often be swamped by immigration and changing behaviour patterns which determine choice of marriage partner. Selection may indeed be at work adapting us to the new urban environment we have created, but the prospect of evaluating it in the shifting kaleidoscope of gene arrays would appear slim unless the effort devoted to such studies is vastly increased.

Of the recognized processes which can influence the genetic make-up of human populations, the most important would seem to be the progressive reduction in the opportunity for natural selection, the break-up of isolates due to improved communications, and the probable increasing scope for assortative mating, especially with respect to educational attainment and general interests. It is difficult to predict the long-term consequences of the relaxation of selection, since we do not know the scope and intensity of stabilizing or directional selection either at present or in the recent past. An increase in the frequency of genotypes which would formerly have been discriminated against appears likely, and this would tend to greater variability. Accentuation of assortative mating would also promote greater genetic variation, so from these two kinds of change we might expect a general enhancement of phenotypic diversity in the population.

Such considerations lead naturally to the question whether man should intervene to influence gene frequencies by eugenic measures. It should be obvious from the foregoing discussion that our ignorance of the genetic control of development and the nature of the inheritance of human traits is so great that direct intervention might seem at best ineffective, and at worst foolhardy. Some commentators, especially H. J. Muller, have been so concerned about our increasing load of mutations and other deleterious genetic changes which have hitherto been kept in check by selection, that intervention has been advocated. He also felt that the problems facing mankind call for greater intellectual capacity and higher degrees of altruism, which might be achieved by setting up sperm or egg banks from suitably distinguished persons who might then act as parents by artificial insemination or implantation, even after they were dead. This might seem analogous to the practice of the animal breeder who uses semen from a superior bull to serve a great many cows. The important difference, however, is that the sensible animal breeder judges the bull by his progeny, not by the bull's own performance, so the proposed step is more akin to mass selection on phenotype which, even with moderate levels of heritability, would in time produce a change in the mean. But for traits so highly valued by would-be improvers of the human lot, such as altruism and intellectual or artistic ability, who would hazard a guess at the degrees of resemblance between parent and offspring due to genetic causes? With wider genetic understanding in the population as a whole, it is conceivable that social organization might be so arranged as to favour higher fertility on the part of those who showed outstanding service to the community, or who were distinguished in some socially desirable respect, so there would be an in-built eugenic trend. To the sceptic, to achieve a degree of consistency long enough to have much genetic effect might seem unlikely. But, even if such measures were successful, the future of man is so imponderable that the genetic changes might turn out not to be what was needed at some time in the future. As a species, man is on the horns of a dilemma if he sets out to do anything about his own genetic make-up which involves directional change.

If we hesitate to intervene in a positive way, a better policy might be to aim at preventing deterioration of the present genetic condition. That would entail a regular check on disease and ill health due to newly arisen mutational change, determining the frequency and regional distribution of genetically determined diseases, and searching for changes which might be related to the relaxation of natural selection. Hence, if mankind is prepared to look to the future and not disguise the fact that we may be moving into a new situation with novel consequences, it will be necessary to establish fully comprehensive records for detailed and continued analysis and comparison.

Of course other possibilities of coping with genetic disease and abnormalities may be contemplated, such as more effective manipulation of growth and development and improved "medicare", or more fundamental intervention at the primary level, i.e. by what is commonly called "genetic engineering". Progress in these fields is so rapid that, given continued stability of industrial society, few would dare to predict what might be possible in 50–100 years. Nevertheless, the complexity of human inheritance is so great that such intervention would appear chiefly applicable to monogenically determined conditions, and this would not remove the need for extensive monitoring of our species with respect to possible changes in the direction and intensity of selection. But more important than the genetic effects are the ethical and political consequences of such intervention for traditional values of the worth and integrity of the individual—an aspect to be kept firmly in mind whenever any proposal for direct intervention is being considered.

Since the future of mankind is so uncertain, it is perhaps salutary to reflect that contemporary discussions of genetic risks and possibilities may look like absurd complacency to our successors. The current trend to relaxation of natural selection may turn out to be of brief duration. Social upheavals generated by population pressure and the conceivable breakdown of industrial society may re-establish the opportunity for natural selection in new directions and at more traditional levels.

FURTHER READING

Allison, A. C. (1954), *Trans. Roy Soc. Trop. Med. Hyg.*, **48**, 312.

Angel, J. L. (1940), *J. Gerontol.*, **2**, 18.

Bajema, C. J., Editor (1971), *Natural Selection in Human Populations*, John Wiley & Sons Inc., New York, London, Sydney, Toronto, ix+406pp.
    A collection of papers dealing with different aspects of the evidence for on-going evolution in contemporary societies. They vary greatly in their suitability for the general reader, but the book is well supplied with illustrative tables and graphs.

Baker, P. T. and Weiner, J. S. (1966), *The Biology of Human Adaptability*, OUP, vii+541pp.
    A little less recent than the other texts, but worth reading for some of the primary data on human adaptation and the comparison of ethnic groups.

Beckman, L. (1959), *Hereditas*, **45**, 1.

Bielicki, T. and Welon, Z. (1975), in *Natural Selection in Human Populations*, Ed. C. J. Bajema, John Wiley & Sons Inc., N. Y.

Bodmer, W. F. and Cavalli-Sforza, L. L. (1971), *Genetics, Evolution and Man*, W. H. Freeman, xi + 782pp.
An authorative account of population genetics in relation to man; presented in a lucid style. Anyone wanting to come to grips with the problem of natural selection in man should read this book, as well as Harrison's.

Clarke, C. A. (1961). *Progress in Medical Genetics* (I), Grune & Stratton, N. Y.

Crow, J. F. (1958), *Human Biol.*, **30**, 1.

Crow, J. F. (1971), in *Natural Selection in Human Populations*, Ed. C. J. Bajema, John Wiley & Sons Inc., N. Y.

Damon, A. and Thomas, R. (1971), in *Natural Selection in Human Populations*, Ed. C. J. Bajema, John Wiley & Sons Inc., N.Y.

Harrison, G. A., Weiner, J. S., Tanner, J. M. and Barnicot, N. A. (1977), *Human Biology: An Introduction to Human Evolution, Variation, Growth and Ecology*, Oxford University Press. 2nd edition, xiv + 499pp.
An up-to-date survey of human variation; contains an immense amount of information and is easily the best account of the subject.

James, W. H. (1970), *Population Studies*, **24**:(2), 241.

Karn, M. N. and Penrose, L. S. (1951), *Ann. Eugen.*, **16**, 365.

Kirk, D. (1968), *Proc. Nat. Acad. Sci.* (USA), **59**, 662.

Lerner, I. M. and Libby, W. S. (1976), *Heredity, Evolution and Society*, 2nd ed., W. H. Freeman & Co., San Francisco.

Medawar, P. B. (1971), in *Natural Selection in Human Populations*, Ed. C. J. Bajema, John Wiley & Sons Inc., N. Y.

Muller, H. J. (1965), in *The Control of Human Heredity and Evolution*, Ed. T. M. Sonneborn, Macmillan Co., N. Y.

Neel, J. V. (1975), in *Role of Natural Selection in Human Evolution*, Ed. F. M. Salzano, North Holland Publ. Co., Amsterdam.

Roberts, D. F. (1953), *Amer. J. Phys. Anthrop.*, **11**, 533.

Salzano, F. M., Editor (1975), *The Role of Natural Selection in Human Evolution*, North Holland Publ. Co. Amsterdam, Oxford, xiii + 439pp.
A multi-author work which provides a comprehensive discussion of the chief concepts. It assumes familiarity with basic genetics and statistics.

Shapiro, H. L. (1939), *Migration and Environment: A study of the physical characteristics of Japanese immigrants to Hawaii*, O.U.P., London.

van Valen, L. and Mallin, G. W. (1971), in *Natural Selection in Human Populations*, Ed. C. J. Bajema, John Wiley & Sons Inc., N. Y.

Waddington, C. H. (1953), *Evolution VII*, 118.

Weiner, J. S. (1971), *Man's Natural History*, Weidenfeld & Nicolson, London, xii + 254pp. A very readable survey of the evolution of man to the present time, set in an ecological framework.

Wyndham, C. H. (1965), *S. Afr. J. Sci.*, **61**, 11.

CHAPTER FOUR

IS MAN A MACHINE?

JOHN LENIHAN

MAN'S PLACE IN CREATION CAN BE DESCRIBED AND STUDIED IN RELATION TO many different environments. The cosmic environment gives us heat, light and gravitation. The geological environment provides other basic necessities for life, including air, water and soil. The chemical environment, still largely natural, though increasingly artificial, provides the materials for the cyclic transformations which, in the biological environment, regulate the transient configurations of ions and molecules that we know as living things.

Though we do not fully understand the internal economy of the human body, vigorous (and sometimes successful) efforts are being made to create an artificial environment, by developing machines and devices to augment or replace various systems of the body. The natural environment, which we are now trying to imitate, is difficult to study. Though the gap between form and function has, in principle, been bridged by the recent achievements of molecular biology, the transformation of biology into an exact science is still a remote prospect.

Models of man

The breadth and depth of this task can be illuminated by examining the concept of man as a machine. This concept embodies the scientific aspect of medicine. Science advances by making models (or theories or hypotheses) which, by the use of technology (in the form of experimental apparatus and measuring systems), can be tested against reality and, if necessary, further refined to provide closer approximations.

The models made by scientists (and by physicians, to the extent that they are scientists) are based on technology and are often most effective when based on the latest available technology. Not surprisingly, the concept of man as a machine began to flourish in the seventeenth century, at the time when technology was providing the inspiration of modern science. The early models of man were, like the technology of the day, based on mechanics. Harvey saw the heart as a pump and the circulation as a hydraulic circuit. Descartes, a little later, had no doubt that the body was a machine; he also used the hydraulic model:

You may have seen in the grottos and fountains which are in our royal gardens that the simple force with which the water moves in issuing from its source is sufficient to put into motion various machines and even to set various instruments playing or to make them pronounce words according to the varied disposition of the tubes which convey the water. And indeed one may very well compare the nerves of the machine which I am describing with the tubes of the machines of these fountains, the muscles and tendons of the machine with the various engines and springs which serve to move these machines...

Although Descartes asserted that the body was a machine, he considered the machine as a passive device, wholly dependent on instructions from a controller:

When the rational soul resides in this machine, it has its principal seat in the brain and may be compared to the fountaineer who has to take his place in the reservoir when all the various tubes of these machines proceed whenever he wishes to set them going, to stop them or in any way to change them.

Other bioengineers were quick to exploit advances in technology during the seventeenth and eighteenth centuries. Borelli (1608–79) made mechanical models of muscular effort representing the arm, for example, as a system of levers and calculating tensions in muscles. His contemporary Baglivi (1669–1707) outlined a course of study for his students:

As to its natural actions, it is truly nothing but a complex of chymicomechanical motions, depending upon such principles as are purely mathematical. For whoever takes an attentive view of its fabric, he will really meet with shears in the jaw bones and teeth ... hydraulic tubes in the veins and arteries, a pair of bellows in the lungs, the power of a lever in the muscles, pulleys in the corners of the eyes, and so on.

Richard Mead, a London physician whose patients included Queen Anne and Isaac Newton, was fascinated by the new science of mechanics and advocated the teaching of mathematics to medical students as an aid to better diagnosis and treatment—but also because

it is very evident that all other methods of improving medicine have been found ineffective these three or four thousand years.

Then, as now, the bioengineers were constrained by the limits of technology. Harvey (1578–1657) could deal with pipes and pumps. His book on the circulation (*de Motu Cordis*) was a great scientific achievement. But his later work on the reproductive system (*de Generatione Animalium*) having no base in technology, was a philosophical essay, Aristotelian in outlook.

Models of the material environment were abundant in the eighteenth and nineteenth centuries. In Glasgow, Joseph Black and Adair Crawford recognized, towards the end of the eighteenth century, the similarities of respiration and combustion—an insight which later grew into the model of man as a heat engine and, even more usefully, into the thermodynamic model (based on the conversion of energy from one form to another) which to a large extent subsumes the chemical, biochemical and physiological models.

Individual organs and systems in the body were also modelled in mechanical terms. At the beginning of the nineteenth century textbooks described the eye as a telescope. Later it became a camera obscura, then a simple box camera; now we recognize it as a very sophisticated camera, with the original zoom lens and through-the-lens exposure control, giving instant three-dimensional colour pictures on everlasting film. The ear was described by Helmholtz in 1868 as a harp. Later it became a telephone, and now it is apparently a miniature hi-fi stereophonic system with contrast compression and other ingenious features. The brain was once a telephone exchange, then a computer, and more recently a holographic data storage system.

In this progression of ideas we see models which are at first wholly descriptive (like Descartes' fountains) then analytical (like Borelli's levers) and finally inductive, leading to general principles and practical applications; the chemical model of man was the beginning of the science of nutrition and the biochemical model provided the basis for the development of new drugs.

The practice of medicine, which was once Hippocratic—based on the study of the whole patient in relation to his various environments—is now largely Cartesian, based on the concept of man as a machine. The task of the physician (and of the scientists who are increasingly important in the clinical realm) is to identify faulty components or systems, and to repair or replace them.

The Cartesian style of diagnosis and treatment is closely related to the two twentieth-century technological revolutions in medicine. The appearance in 1899 of aspirin, the first factory-produced synthetic drug (if we exclude anaesthetics such as ether and chloroform) marked the beginning of rational chemotherapy, subsequently characterized by sulphonamides, antibiotics, analgesics, tranquilizers and many other drugs which have greatly altered the pattern of disease and the quality of life. The chemotherapeutic revolution has now stabilized, with old drugs disappearing from the market as new ones appear.

Influence of electronics

Since 1950 medicine has been invaded by new technology based on electronics. The significant event in this revolution was the invention of the transistor in 1948. Electronics was a thriving activity before then, but its valves and batteries were (because of their size, power consumption and heat production) incompatible with the internal environment of man. The transistor has been exploited for tasks such as satellite communication systems, space research and control of nuclear reactors, in which sophisticated miniaturized equipment has to work with extreme reliability and little power consumption in unfriendly environments for long periods of time. These, of course, are basically the requirements of medicine for equipment to operate in (or closely connected to) the human body.

The exuberance of present-day technology gives ample scope for model making. The conversion of the models from thought experiments into hardware is an exercise which is both more difficult and more important. The abundant technology which inspires bioengineering has, by giving man better control over the environment, made the bioengineer's task much more complicated.

Until quite recently the mechanical improvement of man was limited to external aids. Spectacles, hearing aids and artificial limbs are attached to the outside of the body and do not belong to it in a biological sense. The cardiac pacemaker (page 85) though apparently inside the body is merely an electrical generator connected by wires to the heart muscle. It does not take part in blood transport, chemical processing or any of the other vital functions.

The integration of a machine with the internal economy of the body is peculiarly difficult for two reasons, related to materials and to design strategy. An engineer making a new device has a wide choice of natural and artificial materials, such as wood, metals, plastics, ceramics and semiconductors. The tissues of the body are built from only four basic materials—grit, glue, jelly and soup. Bone is composed of a hard mineral (related to calcium phosphate) and of collagen, which is literally glue; before the development of synthetic adhesives, every large town had a factory, conspicuous by its smell, in which animal bones were boiled to extract the collagen as a source of glue. Muscle, liver, brain and many other tissues are essentially jellies. Blood and other body fluids are salty solutions with suspended proteins, i.e. soups. No engineer could make a computer, a pump or a chemical factory out of such unlikely materials.

Unfortunately the body does not always take kindly to materials that engineers do know how to use.

A system of machines

The engineer is also out of his depth when he tries to imitate—or even to understand—a design philosophy apparently compounded of ingenuity, boldness and incompetence. The body is essentially a system of machines, converting chemical energy (from food) into mechanical energy and heat. The basic process is strangely inefficient. The air that we breathe contains about 21% oxygen, but the exhaled air contains nearly 17% oxygen. In other words more than three quarters of the fuel goes straight into the exhaust. This aspect of the body's design is not as careless as it seems. Oxygen is carried around the body by haemoglobin in the red blood cells. To carry more would mean increasing the number of red cells. The blood would then be more viscous, requiring an increase in the size of the veins and arteries, and a bigger heart to pump the blood. The lungs would also have to be larger to absorb more oxygen. Consequently the body would be bigger, needing more food, and the improvement in fuel efficiency would not be reflected in better overall performance.

The kidney

Other aspects of design are surprising—if not frightening—in their ingenuity. The kidney is essentially a chemical purification plant. The blood, as the body's working fluid, becomes loaded with impurities and must be cleaned up for continued use. A chemical engineer faced with a comparable problem would design a system to remove the unwanted materials by filtration, precipitation and other conventional procedures, leaving the working fluid in a suitable state for topping up and recirculation. The kidney does the job in a very different way. Blood passes first into the glomeruli, which are essentially filter beds. Except for the blood cells and circulating proteins, almost everything is filtered out—vital materials as well as impurities. At this stage the filtrate is, strictly speaking, outside the body and in direct communication (by way of the ureter, bladder and urethra) with the external world. On their way out, the materials needed to sustain life are reabsorbed, leaving only the impurities and a relatively small amount of water to be rejected in the urine. In this way the kidney monitors and maintains at optimum levels the concentrations of about twenty substances in the blood.

An engineer would never design such a machine, since he would rightly be afraid that some malfunction would cause the working fluid to be irretrievably lost. The artificial kidney, as we shall see (page 87) is a very complicated device, less elegant and less adventurous than the natural system.

The heart

The heart, viewed as an engineering exercise, exhibits both good and bad design. The flow of blood through the heart and the blood vessels is streamlined, i.e. free from turbulence. This is not surprising, since the tissues of the growing heart shape themselves around the flowing blood in foetal life, demonstrating a manufacturing technique beyond the capability (almost beyond the imagination) of the engineer. On the other hand, the fuel supply to the heart muscle, the most important muscle in the body, has many odd features. The coronary arteries (so called because they encircle the heart like a coronet) do not appear adequate to their task. The heart, unlike other organs and tissues, does not enjoy a continual supply of oxygen, since the blood flow through the coronary arteries is interrupted by their compression at every contraction of the heart. Another weakness is the lack of oxygen reserve. Most tissues extract 30–35% of the oxygen from arterial blood, so that extra need can, if necessary, be met by increased extraction. The heart muscle takes about 75% of the oxygen from the blood passing through it, leaving very little spare capacity to cope with a reduction of blood flow, e.g. after coronary thrombosis. Worse still, the coronary circulation does not embody the connections, familiar elsewhere in the body, allowing an alternative supply of blood to tissues at risk to the failure of the artery usually supplying them. When a coronary artery is blocked, the tissue that it serves receives no blood from neighbouring arteries, and therefore ceases to function. All of these factors contribute to the prevalence and seriousness of coronary disease in man.

Spare parts

Another feature of the body's design causes almost insuperable difficulties to the surgeon and the engineer. Most machines are made so that faulty parts can be replaced. The body seems suitable for spare-part surgery, since all hearts, lungs and kidneys (for example) are very similar in structure and function. But human engineering embodies the strange concept of mass production, using components which are apparently identical but are not interchangeable. There is, of course, an explanation for this aspect of the design philosophy. Human life is vulnerable to attack by microorganisms of many kinds which, if unchecked, can destroy cells, deprive them of essential nutrients or produce toxic substances which interfere

with the body's chemical processes. Infectious organisms are everywhere
—in air, water, food and in our bodies. Without effective defences against
them, no human being would survive the first days of life. The defence
mechanism depends on a method for recognizing foreign protein. The
lymphocytes which form part of the white-cell population in the blood
constitute a mobile security patrol. Faced by a fragment of protein, such
as an invading microbe, the lymphocytes recognize it (by means which
are not fully understood) as *self* or *not-self*. The protein which passes the
test (e.g. blood cells) evokes no response, but foreign protein stimulates
the mobilization of various defensive mechanisms. A burglar alarm sen-
sitive enough to respond to a single bacterium or virus will naturally be
violently provoked by the invasion of a large contingent of foreign
protein, such as a new heart or a kidney. It is for this reason that organ
transplants present such difficulty. By matching the immunological
characteristics of donor and recipient, kidney transplants can often be
made to survive for several years, but heart transplants have now almost
ceased because in the present state of knowledge they are bound to be
rejected after a relatively short time.

Work on the development of artificial organs has been encouraged by the
difficulty of using naturally available spare parts. One tissue which is not
subject to these difficulties and can therefore be transplanted successfully
is the cornea—a tough layer which forms the transparent window at the
front of the eye. The cornea is important because, apart from its protective
function, it provides a major part of the refractive power of the eye; the
rest, along with the ability to accommodate to different viewing distances,
is provided by the lens. If the cornea becomes opaque, through disease or
injury, vision is lost and can only be restored by a transplant. The cornea
has no blood supply, but obtains oxygen and other nutrients by diffusion from neighbouring tissues. Since the recognition system is carried
by lymphocytes in the circulating blood, the cornea is not vulnerable to
the challenge and rejection processes which inhibit the incorporation of
other foreign protein. Corneal grafting was an uncommon procedure until
the discovery in 1935 that cadaver material could be used successfully.
The eye banks established in many cities during the past thirty years have
been even more successful since the development in the late 1960s of techniques for the long-term storage of corneal tissue by freezing. Corneal
grafting is a relatively simple surgical procedure. A patch (usually circular
or square) is removed from the diseased or injured eye and a similar
patch from the donor eye is stitched in place. Corneal grafts occasionally
fail (by becoming opaque) for reasons which are not fully understood,
but the success rate is high.

Bone is another tissue free from the immune reaction and therefore capable of being transplanted without the danger of rejection. This property arises because the bone tissue is penetrated only by the blood plasma and not by the cells. Incidentally this is the reason why a broken bone takes so long to heal. Lacking contact with red cells, which bring oxygen and take away carbon dioxide very efficiently, bone tissue has to rely on the transport of these gases by the blood plasma which is about 60 times less efficient. Transplanted bone, whether from the patient himself or from a bone bank, does not survive in the recipient's body. However it acts as scaffolding and is gradually replaced by new bone.

Another tissue free from the immune reaction is cartilage, the precursor of bone in foetal and early infant life, which has no blood vessels and is nourished mainly by diffusion. Cartilage is quite widely distributed in the adult body, e.g. in the nose, the pinna of the ear, the trachea and the epiglottis. Cartilage transplants are being widely used—but less frequently since the arrival of silicone rubber and other plastic materials which are well tolerated in the body.

Blood is the most frequently transplanted tissue, largely because its immune reaction is not very specific. For the purpose of transfusion there are four main blood groups (not all of which are incompatible) and it is easy to match donor and recipient.

Skin can be transplanted from one part of the patient's body to another and will survive. Skin from another person is rejected after two or three weeks; skin from an identical (monozygotic) twin is not rejected and is often useful as a temporary expedient.

Artificial organs

Apart from these four tissues, transplantation is still beset by unsolved scientific problems related to the immune reaction. Attention is therefore being directed to the development of mechanical substitutes for body organs and systems. Here, too, there are difficult problems, most of them arising from the incompatibility of engineering design with biological strategy. One of the simplest and most successful aids is the *cardiac pacemaker*.

The heart is essentially a muscle, but is the only muscle in the body with its own timing device. This is a small region on the surface of the left atrium which has some of the properties of both nerve and muscle, and acts rather like an independent nervous system. The natural pacemaker generates electrical impulses, resulting in contraction which spreads rapidly throughout the heart and keeps it beating; the rate of the pacemaker is adjusted (through the action of nerves and hormones) to cope with changes in physical activity or emotional state. In the condition known as *heart block*, the pacemaker impulses do not reach the ventricles, which then beat at a very slow rate, not sufficient to sustain the normal activity of the body. The patient, if he survives, will have a slow pulse and fainting attacks. This difficulty can be overcome in a crude but effective way be delivering small electric shocks to the right ventricle at a rate corresponding to the normal heart beat. Early pacemakers were external to the body, delivering their output by wires passing through the chest wall (an arrangement which could not be maintained for long because of infection) or by induction between two coils, one outside the body and one inside. By 1960 advances in electronics had made possible the construction of implantable pacemakers driven by mercury batteries and connected to the heart by wires stitched to the outside of the ventricle. More recently it has become customary to pass a wire through a vein into the right ventricle where it rests against the wall. Early pacemakers operate at a fixed rate; more recent developments include the *demand pacemaker*, which has an auxiliary electrode detecting the impulses from the patient's natural pacemaker (still functioning, though its output is blocked) and triggering the contractions of the ventricle accordingly. A demand pacemaker responds in the same way as the natural pacemaker to changes in physical or emotional activity. It is estimated that each year about fifty people per million develop heart block, capable of treatment by an artificial pacemaker. The pacemaker is one of the few unqualified successes of bioengineering, allowing patients to lead normal lives. Other attempts to imitate the internal environment have not been so successful.

Given the prevalence of heart disease—the major cause of death in developed countries—and the difficulties of transplantation, the development of an artificial heart is an obvious goal. Great efforts and resources have been committed to this task, particularly in the United States, but the difficulties are formidable. So far there is no way of achieving a power source of adequate output (150 watts would suffice, though the living heart can do better at peak activity) and modest mass, comparable with the 300 grams or so of the heart. Electric motors are too heavy and generate too much heat. Other power sources—nuclear, clockwork, compressed-air, steam and piezo-electrical—have been studied but are all far-removed from practicality. If the power problem could be solved, serious obstacles would remain. So far there is no synthetic material free from the hazard of forming clots in the flowing blood. The attachment of any artificial heart to body tissues presents another intractable problem. A steady pressure of only 20 millimetres of mercury will distort a bone— a process familiar to children who have their teeth straightened by wire braces. Clearly an artificial heart could not be fixed to bones or to the diaphragm. It must be anchored somewhere, but no solution to this promlem is yet in sight.

Though the mechanical heart is at best a remote possibility, replacement of heart valves has been achieved with moderate success. A heart valve consists of two or three flaps of tissue which open and close as necessary, 80 million times a year. Since the heart valve has no blood supply, it is not subject to the immune reaction and can therefore be transplanted. For various reasons the life of a valve transplant is only a few years. The fabrication of valves from other tissues of the body is thought by some surgeons to offer better prospects. When the site of a diseased valve is distorted, making transplantation impossible, an artificial valve can be used. Synthetic materials simulating the dimensions and action of the natural valve have been successful in laboratory tests but lose their elasticity after a few months in the body. A totally different design, such as a ball of plastic material in a wire cage, has been found useful, though again with an effective life of only a few years.

Transplant surgery is limited in scope. Future progress is not easily planned or predicted, because the inherent difficulties are scientific rather than technical. The pace of technology can be accelerated by the provision of more resources, as was done in the development of the atomic bomb and the first landing on the moon; these achievements, great though they were, depended on well-known scientific principles and needed only massive technology for their fulfilment. The surgical technology of transplantation is difficult and demanding, but is within the ability of a large number of surgeons. The major obstacle is the immunological rejection of foreign tissue, a process embodying many unsolved scientific problems; technology can be forced but science cannot.

The design of the body is so economical in size and weight that the replacement of major working parts *in situ* is seldom possible. External machines, relatively clumsy in design and operation, have been developed as replacements for the kidney and for the heart-lung system. In both machines the blood is removed from the body and circulated through devices which perform the necessary operations before returning it.

The artificial kidney machine

The contrast between the internal and external environments is vividly illuminated by considering the design of an artificial kidney machine. The many automatic control systems which operate reliably and effortlessly in the living body require a great amount of equipment when they have to be imitated mechanically. The living kidney depends, for the most part, on two techniques, both of which can be imitated. Blood reaching the kidney is filtered in the glomeruli. This process is helped by the hydrostatic pressure of the arterial blood which forces water, ions and smaller molecules (up to a molecular weight of about 60 000) through the glomerular membrane into the tubules. Pressure-assisted filtration is usually called *ultrafiltration* and is a familiar technique in the chemistry laboratory, where filtration is often accelerated by using a simple water pump to create a partial vacuum in the receiving vessel under the filter paper or disc. Ultrafiltration in the glomeruli removes from the blood the urea and other waste products—but also takes out the essential electrolytes, such as sodium, potassium and chlorine. In the living kidney these materials are reabsorbed from the tubules by diffusion, osmosis and more complicated processes. In the artificial kidney, filtration and reabsorption are combined in a single-stage process, regulated by a semi-permeable membrane. Many plastic materials, when made into thin sheets, act as semi-permeable membranes, allowing small molecules and ions to pass, but holding back larger molecules. The membrane acts partly as a sieve, but the main factor is the more rapid diffusion of small molecules. Cellophane, used in early kidney machines, is a material suitable for the dialysis of blood, holding back the cells and plasma proteins as well as bacteria and viruses, but allowing free passage to the inorganic ions, molecules and the potentially toxic organic substances of relatively low molecular weight, such as urea, uric acid and creatinine. In the artificial kidney, dialysis is achieved by passing the patient's blood over one side of a membrane and passing a dialysing fluid (or dialysate) of appropriate composition over the other side. Materials of small molecular weight will move through the membrane so as to make their concentrations on the two sides more nearly equal. Consequently ions can be moved into or out of the blood by adjusting their concentrations in the dialysate. Toxic substances, not present in the dialysate, will be removed from the blood. Useful materials, if present in the dialysate at the same concentration as in the blood, will be retained.

The dialysing fluid usually contains chlorides of calcium, magnesium, potassium and sodium at the same concentrations as in the blood, or slightly less. The pH of the dialysate is adjusted by adding sodium bicarbonate, or sodium acetate which is readily converted to bicarbonate in the body. Since a patient with kidney failure will produce little or no urine, it is necessary to ensure a net outflow of water from the circulating blood. This is done by adding dextrose (a sugar of high molecular weight) to the dialysing fluid. The dextrose molecules are too large to pass through the semi-permeable membrane, but their presence increases the osmotic pressure of the dialysate and therefore draws water from the blood.

The membrane itself may be made in various forms. In early machines the blood flowed through a tube of cellophane, immersed in a bath of dialysate. In later designs the membrane is in flat sheets, the blood flowing over one side and the dialysate over the other. The membrane must have adequate mechanical strength in thin sheets and a suitable structure for dialysis. These requirements are not too difficult to meet, but the membrane must not adhere to platelets or other constituents of the blood, and must not release toxic substances into the blood.

There remain two problems: the delivery of the patient's blood to the dialyser and the preparation of the dialysing fluid. The blood for laboratory tests is usually obtained from a vein, i.e. from the low-pressure end of the system. For satisfactory external circulation it is better to tap the high-pressure end by opening an artery, returning the purified blood to a vein. The difference between arterial and venous blood pressure is, in some designs of artificial kidney, enough to drive the blood through the external circuit, so avoiding the damage to red cells which is inevitable when a pump is used. If artificial kidney treatment is needed only once, e.g. to deal with an acute poisoning episode, the connection of the machine to the patient is not unduly difficult, using the conventional method with a cannula (i.e. a tube) in a convenient artery of the arm or leg; the returning blood passes through another cannula into a vein. A blood vessel may safely be cannulated for a day or so, but after that becomes vulnerable to infection, thrombosis and other complications. The first successful means of achieving long-term connection between the internal and external environments was the Scribner shunt, a plastic tube connecting an artery or vein in the wrist or lower leg. The shunt was removed to allow connection of the patient to the machine and replaced at the end of the treatment. It is now more common to use an arterio-venous fistula, i.e. a permanent surgically created connection under the skin between an artery and a vein. Access to the fistula is obtained by a needle.

The preparation of the dialysing fluid is unexpectedly difficult. The solution of minerals and sugar, warmed as it must be to body temperature, could not be kept sterile in the amounts needed—400 litres per session or even more. So the fluid is stored in concentrated form to prevent bacterial growth, since the enormous osmotic pressure would destroy microorganisms by drawing off their water content. The fluid is diluted just before it is used.

It would clearly be attractive to recirculate the dialysing fluid. This is done in some systems, especially when portability is desired, by using a succession of chemicals to remove unwanted materials from the fluid. More usually the dialysing fluid is used once and discarded, since in this way higher concentration gradients and therefore more effective dialysis can be maintained. Dialysis of the patient's blood in the external circuit can be augmented by ultrafiltration. Using a pump in the dialysate outflow line and a flow-retarding valve in the inflow, the dialysate can be delivered to the membrane at a negative pressure, so encouraging the extraction of water from the blood; small molecules and ions are dragged through with the water. An alternative method is to increase the hydrostatic pressure of the blood delivered to the dialysis chamber. Ultrafiltration ensures that, in the event of a leak in the dialysis membrane, blood will pass into the dialysate where it can be detected by a photoelectric cell responding to change in colour. Without the hydrostatic pressure gradient, it would be possible for dialysing fluid to leak into the blood. This is a more serious hazard, partly because the leakage is not so easily detected but mainly because, despite all the precautions that are taken, the dialysate may not be completely sterile.

A number of other precautions are necessary. To prevent coagulation in the external circuit, the blood must be continuously infused with heparin as it leaves the patient. The effect of the heparin is neutralized by infusion of protamine before the blood is returned to the patient's vein. It is also important to fit a bubble trap in the tube returning blood from the dialyser to the patient; gas bubbles are not a hazard in the external circuit but are dangerous inside the body, because they can cause blockage of blood vessels.

The artificial kidney is a bulky machine and its performance does not match that of the living organ. Since permanent connection of the patient to the machine is so irksome, patients with chronic kidney failure are usually dialysed intermittently at a high rate. The healthy kidneys produce an average 1 ml of urine per minute. The concentration of urea in the blood is then maintained at about 40 milligrams per 100 ml; urea is a product of protein metabolism and is the most important waste material excreted in the urine. However, a patient can tolerate blood urea levels up to about 400 milligrams per 100 ml and, on a low-protein diet can live for a few days without severe toxic symptoms. Consequently dialysis twice a week will keep him reasonably healthy. It is therefore possible for dialysis to be undertaken in the patient's home, allowing him to lead a reasonably normal life in the intervals between treatment.

The cost of artificial renal dialysis is high—at present (1978) about £10 000 per year in hospital and more than £12 000 per year at home. The cost of the machine (£2000 to £4000) is not a major part of the expense; much well-meant effort is spent by charitable organizations in raising money to buy kidney machines for hospitals which already have them in store, unused through lack of funds or expert staff to operate them. It is seldom appreciated that renal dialysis needs considerable medical, nursing and laboratory resources. Unfortunately there are no realistic prospects of dramatic improvements in design or performance. Well-publicized projects to produce pocket-size artificial kidneys have all failed—and will continue to fail—because our present skills, techniques, materials and understanding allow us to produce only a very crude imitation of the natural system.

The heart-lung machine

The interior of the heart was the last region of the body to become accessible to the surgeon. The problem is difficult. Obviously the surgeon cannot operate on a heart that is still beating and therefore moving. But if the heart stops beating, the patient dies in a matter of seconds. The solution involves a device which by-passes the heart and lungs, keeping the rest of the body oxygenated while the surgeon operates on the valves, coronary arteries or other tissues in the heart.

As an imitation of natural function, the heart-lung machine is even less successful than the artificial kidney, since it cannot be used in a conscious patient and, even in the anaesthetized state, will support life for only three or four hours.

The heart-lung machine has to perform two functions: pumping (normally done by the heart) and gas transfer (normally done by the lungs) involving the addition of oxygen and the removal of carbon dioxide. It might be thought necessary only to take over the pumping action of the heart, leaving the lungs to deal with gas exchange. In practice this would mean a more complicated machine, with twice as many pipes and more hazard of leakage, blockage or other misfortune. Use of the heart-lung machine allows the lungs to be collapsed during the operation, giving the surgeon more room.

The pumping action of the heart can be imitated mechanically without undue difficulty, but the blood is inevitably damaged. Red cells may be crushed by the pump or burst open by cavitation—the momentary fall in pressure which occurs with turbulent flow. The loss of red cells is not in itself serious, but the disposal of the haemoglobin which is released puts an additional strain on the kidneys.

The necessary gas exchange can be achieved by bringing the blood into contact with oxygen, which then replaces the carbon dioxide which has been produced by metabolic activity in the tissues of the body. In one commonly used system, oxygen is bubbled through the blood in the external circuit. This is quite an efficient method, though removal of the bubbles requires care. In another design, rotating stainless-steel discs dip into a pool of venous blood and carry it into an atmosphere of oxygen. Yet another possibility is the screen oxygenator, in which venous blood flows over plastic-coated metal gauze in an atmosphere of 100% oxygen.

The most recent design simulates the natural lung a little more closely by passing the blood over a thin plastic membrane with oxygen on the other side. The membrane oxygenator has many attractions, along with a few disadvantages. Its internal structure is so complicated that it cannot be cleaned after use and therefore has to be disposed of, adding to the cost of the operation. Currently available membranes are liable to damage the blood proteins and to adhere to platelets—the danger which occurs also in artificial kidney machines. Membrane oxygenators are, however, being used successfully in some hospitals.

The heart-lung machine has led to important advances in surgery and has given new life to many patients. But the machine does not begin to approach the efficiency or elegance of the living system that it imitates. Here, as in other directions, some of which have been indicated in this chapter, bioengineering appears as one of the less successful adventures of modern technology. Yet the imitation of the biological environment is, in the long term, one of the greatest challenges facing man's ingenuity and understanding.

Man's increasing power to command his external environment is inconsistent with the slow pace of biological evolution. As J. D. Bernal wrote half a century ago:

Normal man is an evolutionary deadend; mechanical man, apparently a break in organic evolution, is actually more in the true tradition of a further evolution.

Bernal, looking far into the future, saw the limbs and the digestive system as parasitic appendages, doing nothing which could not be done better by mechanical or chemical technology. He saw future man as a combination of living organs and machines, specialized for intellectual development and for remote sensing and manipulation of the external environment.

Our limited success in bridging the gap between the internal and external environments of man reminds us that progress is not inevitable, and that there is not necessarily a solution to every problem. The attack on this particular problem may not be successful until engineers forget their justifiable pride in the achievements of technology and adopt the humbler attitudes of the biologist.

FURTHER READING

Bernal, J. D. (1929), *The Worlds, the Flesh and the Devil*, 2nd ed. 1970, Cape, London.
Longmore, Donald (1969), *Machines in Medicine*, Aldus Books, London.
Longmore, Donald (1968), *Spare-part Surgery*, Aldus Books, London.
Longmore, Donald (1971), *The Heart*, Weidenfeld and Nicolson, London.
Najarian, J. S. and Simmons, R. L. (1972), *Transplantation*, Lea and Febiger, Philadelphia.
Nossal, G. J. V. (1971), *Antibodies and Immunity*, Penguin Books, London.
Rapaport, F. T. and Dausset, J. (1968), *Human Transplantation*, Grune & Stratton, New York & London.

CHAPTER FIVE

MICROBES AND THE ENVIRONMENT

JOHN E. SMITH

Introduction

It is now nearly 300 years since Anton van Leeuwenhoek became the first man to witness the amazing living world that exists beyond the limits of our vision. Leeuwenhoek was a man so possessed by the fascination of grinding lenses and magnifying objects that he meticulously devised his own primitive but remarkably successful microscope. He wrote hundreds of letters to the Royal Society in London in which he tells of his discoveries with his "magic eye": exploring the mysteries of the eye of a fly, the leg of a louse and many other formerly unseen details of life. However, it was to be his examination of "clear" rain water that was to serve as the foundation of a new branch of science—microbiology. This was the first time anyone had seen the submicroscopic world populated by teeming millions of living microscopic forms—his "wretched beasties", so small that "this last kind of animal is a thousand times smaller than the eye of a large louse". Leeuwenhoek thus became the first of the microbe hunters and, in the years to follow, others including Spallanzani, Pasteur and Koch sought the nature of these organisms and demonstrated their causal involvement in disease and decay.

Microbes are a varied and complicated group of forms comprising viruses, bacteria, fungi, algae and protozoa. Many of these are submicroscopic, while others come within our limits of vision because of their colonial nature of growth. Linnaeus, the great classifier of living creatures, palled at bringing order to this group. "They are too small, too confused, no one will ever know anything exact about them, we will simply put them in the class of Chaos."

93

Modern microbiology, however, is a very exact science and much is known about the microbes and the important role they play in the ecosystem. Microbes exist throughout the biosphere and, because of their vast numbers and unique biochemical mechanisms, they exert a major influence on man's existence. This chapter highlights a few of the intriguing relationships between man, his environment and the microbes.

Environmental decomposition

When organisms die they ultimately fall to the soil or sink to the bottom muds of rivers, lakes or oceans where they are eventually decomposed by other organisms—particularly the microbes that exist in these environments. Such microbes are decomposers or saprobes living on dead organic matter, and as such are nature's scavengers. The dead organisms, or indeed any other organic material (e.g. faeces, pesticides, litter) entering these environments, will be gradually disintegrated structurally and chemically into simple components. Decomposition is a complex interaction of physical and biological agencies. Thus the valuable components of biological material are released from complex combinations and returned to the ecosystem where they can be re-used by other forms of life (figure 5.1). Without such recycling, there could be no continuation of life.

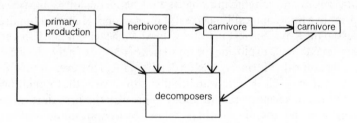

Figure 5.1 The cycling of nutrients through an ecosystem.

The decomposing or decay-promoting activities of microbes are varied and enormous. Microbes exist in all environments in active or dormant forms and, because of their enzymatic diversity, they can colonize almost all organic substances.

Natural and applied decomposition

Soil
The importance of soil as a vehicle for crop production has long been recognized by agrarian societies, and modern agricultural practices are

dependent on maximizing soil fertility. Soil is a complex heterogeneous system, both chemically and physically, and is composed largely of mineral particles and organic material adhering to form aggregates of differing sizes and stabilities. Soil abounds with different forms of microbes. Typically, the microbial population is large and complex, but not always active. Bacterial counts of the order of 10^6–10^8 per gram of soil are quite common and the biomass of bacteria in agricultural soils can be 1600–35 000 kg/ha. The growth of microbes in soil is subject to great restrictions, such as local scarcity of nutrients and availability of free water. Soil is not an ideal environment for microbial growth but rather should be viewed as an environment that permits the survival of a varied population of opportunists with immense biochemical potential. Many microbes will exist in resistant spore forms but can rapidly reactivate and utilize any organic material that should arrive in their microenvironment. After the feast, the microbes will largely resume their quiescent form.

How important are microbes to soil fertility, apart from releasing bound materials from complex forms? The ability of some bacteria to fix atmospheric nitrogen into organic form has benefited agricultural productivity since farming began. Many of these microbes, such as *Azotobacter*, are free-living and are present in most soils. Others, such as the *Rhizobia*, can be free-living and can also enter into the roots of leguminous plants and establish a stable relationship, gaining organic material from the host plant and in return giving nitrogen compounds beneficial to the host. The possibility of transferring the genetic information for N-fixation from the bacterial genome to the roots of plants other than the pea family is being earnestly pursued. If this could be achieved, the effects on world agriculture and food availability would be immense, since current costs of nitrogen fertilizers are out of reach of many developing nations.

Microbial populations are generally highest around the growing roots of plants and trees, due primarily to organic exudates, and they constitute the rhizosphere population. As many as 10^{10} microbes/g root surface soil can be found, with the population decreasing with distance from the root. Such microbial populations can have beneficial and detrimental effects on plant growth. In extreme cases, pathogenic microbes enter and colonize the host root and subsequently kill the plant. In other cases, intensive growth of harmless microbes can exclude pathogenic organisms. The dynamic state of these microbial interactions is being extensively studied. There is evidence that some microbial combinations are definitely beneficial to plant growth and may perhaps be artificially stimulated by soil inoculation.

Sanitation and pollution control

The domestic and industrial activities of man have led to an ever-increasing volume of organic materials entering the environment. Faeces represent man's largest solid contribution to the environment and, although the microbes (approximately a quarter of the mass of faeces) are mostly harmless, they can be sources of diseases such as cholera and dysentery. Many diseases are effectively transmitted from faeces to mouth by way of drinking water. One of the main factors contributing to the control of serious water-borne epidemic diseases in modern times has been the development of adequate sewage disposal and the supplying of purified water. Modern sewage disposal is fundamentally a controlled fermentation process in which a broad spectrum of aerobic bacteria, fungi and protozoa inactivate pathogenic microbes and break down the organic wastes into harmless components that can be safely returned to the environment.

Whereas in bygone days water was obtained principally from wells (often contaminated by faecal organisms), nowadays in developed communities water is collected in huge reservoirs where it is purified and chlorinated before being distributed to the populations of large areas. Modern society is totally dependent on a supply of pure water and the efficient disposal of sewage. When such services break down, under conditions of war or natural disaster, disease can rise rapidly to epidemic levels. Regrettably most citizens are ignorant of the valuable role played in our society by the sanitary engineer and public health microbiologist.

Industrial organic wastes are also disposed of by microbial breakdown in conventional sewage systems. However, many industrial concerns treat waterways and oceans as a gigantic sewerage system and indiscriminately discharge untreated wastes into these systems. Many of our rivers and inland water systems have already been overtaxed, and the endemic beneficial microbes have been largely overcome by putrefying bacteria which can only partially break down organic wastes, leading to obnoxious smells and polluted water. Modern man must not destroy his heritage, and more legislation is required to prevent self-interested nations from pumping waste into our environment. Pollution does not recognize international limits.

An increasing pollutant in the world's oceans is oil. The amount of oil entering the seas each year from tankers, oil seepage and industrial wastes has been estimated at over 2 million tons and excludes the massive contribution by aerial fall-out of unburned hydrocarbon and combustion products. Recent years have seen a great increase in the input of oil into the marine environment, but, gross pollution of the sea surface and shores is not yet evident. In major part this is due to microbial attack and the oil's subsequent breakdown into harmless components. Bacteria are by far the most important microbes involved in oil (and indeed any other organic) breakdown in the sea.

Biodeterioration

Man's industrial progress has always been challenged by the decay activities of microbes. However, there was no real awakening to the immensity of this problem in an industrial context until the end of the Second World War. Studies initiated during hostilities indicated the magnitude of the economic losses and the unbelievable range of materials and commodities that could be attacked, damaged or destroyed. The deleterious effect of microbes in the industrial environment has been called *biodeterioration* and can be considered as any undesirable change in the properties of a material caused by the vital activities of organisms. In Britain alone, the annual losses can be counted in hundreds of millions of pounds—a loss that no country can afford. Biodeterioration is a global problem. In the tropics, where high humidity and temperature prevail, it reaches a peak of activity, and many commodities which would be free of biodeterioration in temperate zones will readily succumb there.

How does biodeterioration make itself evident? All industries concerned with food, beverages, paint, toiletries, pharmaceuticals, rubber, plastics, chemicals, paper, wood, textiles and all aspects of engineering will undoubtedly experience at some time biodeterioration problems in the production and/or storage of materials. A few of the many distinctive problems arising are discoloration, putrefaction and quality imperfections in products, breakdown and loss of viscosity of emulsions, coagulation of dispersions, development of unpleasant odours either in the product or product line, rapid corrosion of metal products, blockage and overheating of production machines, and collapse of wooden and concrete structures. In the service industries, biodeterioration can be seen in the corrosion of pipelines and cables, rotting of telegraph poles, malfunctioning equipment, and the physical breakdown of stone surfaces in buildings and works of art. The presence of large populations of bacteria in emulsions and lubricants in engineering works is now recognized as a serious health problem causing both dermatitis and lung infections. The essential nutrients for microbial growth and multiplication are regularly found in the constituents of most manufacturing processes or the product itself. Microbes can cause biodeterioration either by their presence or indirectly by excretion of waste products such as acids which can cause contamination or corrosion. In most types of biodeterioration there is a complex

Table 5.1 Microbial infection of industrial waters

Industry	Most common microbes	Nature of biodeterioration
Paper-making	Slime formers *Alkaligenes, Aerobacter, Pseudomonas, Chlamydobacteriales*	Slime growth occurs where design of equipment encourages laminar hydrodynamic flow rather than turbulence. Slime contaminates sheets.
Gas	Sulphate reducing bacteria *Desulphovibrio*	Occurs during storage of gas in water-sealed gas holders. Bacteria generate H_2S from SO_4 ions in water. Produces 'dirty' gas.
Textile	Slime-producing bacteria	Blockage of spray-nozzles in humidification and air-conditioning plant where large volumes of water are used.
Engineering *(a)* Cutting oils	Many bacteria, sometimes yeasts and fungi	Loss of essential properties. "Cracking" of emulsions causes restricted water flow
(b) Cooling towers		and blockages and loss of cooling effect.
Coal mining	Acid-forming bacteria *Thiobacillus* spp., fungi	Acid corrosion of water extraction pumps and other machinery. Collapse of pit props.
Food	Wide range of bacteria, yeasts and fungi	Decomposition of food. Food poisoning.
Steel	Many bacteria *Thiobacillus, Desulphovibrio*	Acid corrosion, pitting, blockage of pipes, tank corrosion.
Construction	Mostly fungi, *Serpula, Coniophora*	Collapse of wood

ecology of microbes involving many types, all active and interacting continuously. Some common examples of biodeterioration are given in Table 5.1.

The consequences of biodeterioration are expensive to industry and are attracting attention on an international scale. Scientists throughout the world seek to identify and characterize the microorganisms concerned, to study their ecology and relationship with the environment and (most important of all) to develop measures for prevention and cure of biodeterioration. The understanding of biodeterioration offers substantial rewards to industry and to national economies and resources.

Microbial biodeterioration is not limited to industrial situations; it causes vast losses in post-harvest agriculture. All agricultural and horticultural crops are readily attacked by many microbes during the period from harvesting to ultimate consumption. Each year millions of tons of food products are irretrievably damaged and made unfit for human consumption. The critical factor in most cases of agricultural biodeterioration is the high water content of most plant and animal products which makes them highly susceptible to microbial decay.

Agricultural products are produced in an open environment exposed to contamination from microbes from the air and soil, and from handling and transit. Given the right conditions of growth, such microbes will rapidly attack the products. To combat these effects, many countermeasures have been devised:

(1) The rapid consumption of fresh products
(2) When possible, refrigeration or storage under gas to reduce microbial growth
(3) Preservation by drying and addition of antimicrobial chemicals
(4) Pasteurization
(5) Processing and sterilizing by heat and subsequent storage in sterile conditions—jars or cans.

In this way the ravages of the microbe can be limited and needless wastage prevented. However, throughout the world, microbes claim a staggering amount of our agricultural produce.

A new problem of great importance to man and his animals has been the recent observation that highly toxic compounds can be formed in many types of relatively dry products, such as grains, by the metabolic activity of moulds. Because of their relatively low water content, grains and other seed crops have been particularly suitable for storage. Indeed it is believed that the early Nile civilizations grew through their ability to store grain correctly and so leave a proportion of the population to specialize in other crafts and arts.

However, if grain is allowed to become moist, microbial infection will rapidly occur; in extreme examples the piles of grain will become hot and sticky, and will ultimately reach a temperature where spontaneous com-

Table 5.2 Possible routes for mycotoxin contamination of human foods

1. Mould-damaged foodstuffs
   (*a*) Agricultural products, e.g. cereals
   oilseeds
   fruits
   vegetables
   (*b*) Consumer foods

2. Residues in animal tissues and animal products, e.g. meat
   milk
   dairy products

3. Mould-ripened foods, e.g. cheese
   meat products

4. Fermentation products, e.g. microbial proteins
   enzymes
   other food additives

bustion will occur. Such fires are not uncommon in grain terminals; a similar microbial pattern can lead to fires in hay stacks and lofts.

The microbes that attack the drier types of agricultural produce (such as grain) are termed *xerophilic fungi* because of their ability to grow at low moisture levels. Apart from their spoilage effect on grains, many of these fungi are now known to produce toxigenic compounds which can have adverse (and indeed disastrous) effects on man and his animals. Over 100 of these compounds have now been identified and one in particular, aflatoxin, is the most potent hepatic carcinogenic compound known to man (see Volumes 6 and 7). The effect of these compounds has been extensively studied in animals, and many can cause disturbing and serious effects. Such toxic molecules (mycotoxins) are also present in the diet of man, and there can be little doubt that they are already doing damage. The epidemic levels of liver cancer in certain parts of Africa are almost certainly due to the high consumption of mouldy food. Table 5.2 shows the possible routes of entry of these mycotoxins into diet.

Diseases of plants—the relationship to man

Infectious disease of crop plants
The history of man has been dominated by his search for food. Primitive man was by nature and necessity a hunter and food gatherer. An important turning point occurred when he began to domesticate plants and animals. In this way he found a more dependable source of food and began the development of what we now consider as modern agriculture.

By having to spend less effort in search of food, the early agriculturalists found the time to develop crafts and arts, leading ultimately to modern industrial civilization.

Only catastrophic food shortages can make people realize that food production is a complicated and integrated biological process carried on by animals and plants under the supervision of those most practical of biologists—the farmers. The cultivation of crops has always been exposed to the spectacular and damaging effects caused by storms, hail, drought, floods, frost, insects and, of course, microbes in their many forms. Such natural phenomena were viewed by the farming communities as acts of powerful spirits which had to be appeased to ensure a bountiful harvest. Consequently there evolved numerous and varied rituals, traditions, symbols and folklore which all centred around man's desire to increase crop production. The mystical concepts of plant disease prevailed for a long time. It was not until the eighteenth century that European biologists set out to study and control diseases scientifically, and thus paralleled the efforts of medical men who were seeking meaningful reasons for the diseases of man.

We now know that microbial plant diseases are caused by several different types of microbes, including bacteria, viruses and fungi. Whereas, in man and animals, bacteria are the main disease-causing organisms, in plants many more diseases are caused by fungi. It is hard to estimate accurately the monetary losses due to microbial plant diseases. Suffice it to say that in the United States, an agriculturally sophisticated country, there is a 10–14% annual loss in agricultural output, while in less organized and less scientifically aware countries this may well increase to 40–50%. It should be noted that those countries that enjoy the highest standard of living also have the most efficient and high-yielding forms of agriculture.

The science of plant pathology arose primarily to protect plant crops from the ravages of microbes and other deleterious agents such as insects. The plant pathologist aims to control endemic disease by crop hygiene and chemical control in the short term, and by genetical breeding for resistance to attack in the long term. Furthermore, he aims to establish and enforce quarantine measures to prevent the geographical spread of disease, since infectious plant diseases have no respect for frontiers. There is now a complicated programme including advanced research studies carried out in government, industrial and university laboratories which seek to cure and prevent diseases, along with efforts to disseminate this information to farmers. As a field of scientific endeavour, plant pathology is not given the acclaim so readily bestowed on medical research—but civilized man could not exist long without the services of plant pathologists.

The unique biochemical nature of plant species is shown by the almost universal specificity of individual diseases for specific crop plants. Although every species will have numerous individual diseases, few of these diseases affect different species. Indeed, even within species, different strains show quite distinct disease syndromes. Man's predilection for monoculture has undoubtedly been to the benefit of the microbe, for, when correct conditions for disease occurrence and dispersal prevail, there can be explosive outbreaks of individual diseases. In this way catastrophic incidences of plant disease can occur.

When a disease is spread over an area in which the causal agent has been present for a long time, it is termed an *epidemic*. When the pathogen moves into other areas, the disease is called a *progressive epidemic*, and when several continents become invaded it is a *pandemic*. Epidemics and pandemics of plant disease have rampaged throughout the world for as long as historical and scientific records have been maintained. History abounds with examples, and all countries have at some time experienced the indiscriminate attack on agricultural crops by one type of disease organism or another. Fortunately, balanced agricultural policies, involving a range of crops, prevent major suffering. If communities or nations become dependent primarily on one crop, then, when disease strikes a susceptible plant population, untold suffering will occur—disrupting the economic, social and political conditions within that area. In the early 1940s an epidemic fungal disease of rice caused huge losses in production in Bengal. Food shortage, coupled with the political problems of that time and governmental mismanagement, contributed to the deaths of at least 2 million people from starvation and disease.

Most catastrophic diseases have followed a typical pattern: developing as an epidemic, sometimes expanding into a pandemic, and then abating to a condition that can be tolerated in agricultural practice. The two greatest examples of a catastrophic pandemic, the Irish Famine (the late blight disease of potatoes) and the Bengal Famine, covered the entire sequence of events in 2–3 years, whereas a virus disease of trees (peach yellows) lasted a century, and the chestnut blight disease is still in existence after 75 years. The distinguished German plant pathologist Gäuman aptly summed up the situation, stating:

Each epidemic follows its own rules, changes its character, increases and becomes malignant, decreases and becomes milder; it possesses its own physiognomy, its own morphology, its own *genus epidemicus*. In most cases an epidemic has two limiting factors: its beginning and its end are determined by time and space.

The exploration and colonization of the world by man resulted in the movement of potentially useful food plants from one geographical area to another. In this way many new crop plants came to the Western Hemisphere and, as instanced by the potato, became valuable and popular food sources for the expanding population. Pathogens invariably follow their hosts, and it is only a matter of time before the pathogens become endemic in the plant population. The new environment of the pathogen (climatic and soil conditions, and agricultural practice) may also influence the degree of susceptibility of the host.

No one knows why most epidemics and pandemics eventually subside and often become moderate endemic diseases. Most probably the cause is not acquired immunity, but rather selection and elimination of susceptible individuals, and the gradual development of a resistant population. Modern man has in many cases intervened in this cycle of disease by chemically controlling the incidence of the disease organisms. The immediate benefits to our economy are obvious, but the plant remains susceptible to the disease, and treatment must be permanent. When the powdery mildew (*Uncinula necator*) of grapes first reached epidemic proportions in Europe, no effective fungicide was available. Although disastrous damage was done, a resistant plant population arose and nowadays the disease is relatively harmless. In contrast, the downy mildew (*Plasmopara viticola*) of grapes was controlled at the peak of a pandemic with copper compounds (the much acclaimed Bordeaux Mixture) and was never allowed to run its natural course and produce resistant plants. To this date the European vine is still as susceptible to the disease as it was 60 years ago. However, man's economy demands a regular and continuous supply of food and related raw materials, and this can only be achieved by the constant application of all technical advances in disease control.

At present there are many examples, particularly in tropical countries, of plant pathogens that could seriously threaten crops in other parts of the world. Governments must recognize the need to fund studies that will give man information about the factors predisposing the spread of a pathogen, to encourage the international exchange of gene pools for disease resistance and, in particular, to encourage the exchange of trained pathologists able to recognize threatening diseases. Many pathologists foresaw the dangers and potential of the coffee rust disease now such a serious problem in South America. So often these important decisions are made by scientifically unaware administrators who reach decisions from a political rather than a scientific basis. The "green revolution" in agriculture has resulted in major advances in food production in many developing countries. Such increased yields are primarily based on the use of a small core of germ plasm, higher plant density, increased number of crops per year, increased irrigation and other improved cultivation practices. Although all these methods give higher plant yields, they also favour the spread of plant diseases. Are we maintaining a suitable vigilance against threatening diseases in this new form of agriculture? The build-up of human populations on the basis of this new-found food supply is economically, socially and politically hazardous in relation to the possibility of a severe epidemic or pandemic disease. Microbes are always in our environment and need only the slightest opportunity to burst from obscurity to ravage a host population.

A catastrophic plant disease: its impact on man
The natural home of the potato is South America, where it grows wild in many regions. The ancestors of the present-day edible potato were first domesticated far back in antiquity by immigrant Indians on the high plateaus of the Andes. The Spanish conquistadores who first encountered the potato realized its economic and nutritious importance, and soon adopted it as the food for their slaves. Botanical records show that the potato was introduced into Spain about 1560–1590 and soon became an article of commerce. Its cultivation spread rapidly through Europe, reaching Britain towards the end of the century. It was then gradually brought into cultivation throughout Britain and Ireland, and for the next 200 years was not particularly troubled by microbial pathogens.

Coincident with the development of the potato, the Industrial Revolution was evolving in Britain. In order to develop the new industries, the manufacturers needed cheap labour. Cheap labour meant cheap food. The potato was the cheapest and one of the most effective single foods that man has been able to cultivate in the Western Hemisphere, and it gradually came to occupy an increasingly important place in the diet of workers. By the beginning of the eighteenth century it was part of the day-to-day existence of the worker and his family, and throughout the century it was to act as a buffer between the relatively high price of food generally and low wages. The average daily intake of potatoes at that time in the working-class community was 2–4 lb. The potato had a double part to play in the Industrial Revolution—that of a cheap nutritious food, and that of a weapon or tool ready forged for the exploitation of the poor and weak sections of the society. Although the potato came to be cultivated throughout Britain and many parts of Europe, only in Ireland did it tend to become in many areas the sole agricultural crop. In 1845 three quarters of the population relied on the potato for half their food, while the rest relied on it completely. Disaster was soon to strike. The blight first struck in epidemic force in Belgium as a completely new and unknown disease probably arising from South America. The disease visited Britain with calamitous results, nowhere more tragic than in Ireland. For three years the crops were destroyed, and it is impossible to describe the horrors of this period or to compare them with anything that had occurred in Europe since the Black Death of 1348.

How many people died during the Irish Famine will never be known precisely. In 1841 the population of Ireland was 8 174 000. In 1851, after the famine, it had dropped to 6 552 000 and the census commissioners calculated that the normal increase should have been to 9 million—a net loss of 2.5 million people. Between 1846 and 1851 nearly a million people emigrated, and roughly 1.5 million perished during the famine from starvation and diseases brought on by malnutrition.

The most important historical result of the famine was the mass exodus of the Irish. In particular, there was a wholesale emigration to the New World across the Atlantic, thus affecting the destiny of the United States of America. There was, however, another emigration, more numerous though less documented, in which the Irish in ragged hordes crossed the Irish Channel to land at ports in South Wales, England and Scotland. This was the flight of the very poor who could not find the means to pay for the more hazardous Atlantic crossing. Because of their past history and impoverished state, the immigrant Irish became a source of cheap labour. No regulations on wages then existed, and trade unions were still struggling to establish themselves. On many occasions the Irish were used for strike breaking, and as a result it was frequently impossible to get Irish and British labourers to work together. These economic differences were further inflamed by religious discord.

The economic and social upheavals that resulted from this pandemic have never again been witnessed in this country. The disease is still present, but in a much less virulent form. And yet the appalling human suffering did in fact bring about immense and far-reaching changes in the social and economic life of Britain. Coincident with the Irish famine was the fight to abolish the Corn Laws which gave protective tariffs on imported grain. Such Corn Laws were a cruel burden on the poor and the working class. The devastation wrought by the late blight disease forced the suspension and final abolition of the Corn Laws, and speeded the adoption of the principle of free trade. Free trade subsequently made Britain more heavily dependent on foreign food supplies, thus necessitating policies of peace and the development of a powerful navy to protect trade routes.

Trees and society

The forests that cover many parts of our planet are one of man's greatest natural resources. Wood serves to make weapons, ships, homes and furnishing; but perhaps most important of all in a civilized society has been its use as the raw material for making paper and so giving man a means of information dispersal and retention. In the relatively recent past, man has unwisely exploited the resources of forests, but fortunately in most areas a much more rational and realistic attitude has now been adopted. Whereas in earlier times the logging industries would abandon large areas to devastation, and resultant irreparable soil erosion and climatic change, efforts are now made to replant and prepare for future generations.

Microbial diseases may strike at any time, and no tree is ever completely safe. The microbes will eventually win, even if they have to wait a thousand years for a giant *Sequoia*. Young trees are more susceptible; seedlings and saplings will succumb readily to microbial attack through the roots, stems and leaves. However, many survive to become the forests of tomorrow. Forest pathologists are locked in constant contest with the disease organisms, trying to control their incidence and spread by tree hygiene, chemical treatment and long-term genetic breeding.

Fungi, in particular, wreak great havoc with standing trees, attacking not only the commercial timber trees but also the ornamental ones used for landscaping and shade purposes. All trees usually carry some degree of microbial infection, but in most it is contained and never causes severe damage. However, some diseases of trees can be devastating. In many cases mature trees may be 50 to several hundred years old and, when such trees die, their loss cannot easily be replaced by this generation. The Dutch Elm disease (*Ceratocystis ulmi*) is such a catastrophic disease, attacking and destroying mature trees as it relentlessly moves slowly through continents. The disease is due to wound infection and the distribution by man and elm bark beetles of the spores of the causal fungus. The beetles, by carrying the fungus spores into the brood galleries, transmit the pathogen and permit its rapid growth. The disease was first identified in Holland in 1919, spreading rapidly through Central Europe, and in many parts killing all the elms. The disease now exists in America and, more recently, has become a scourge in Britain and is moving northwards. There is little that modern science can do to hold back this army of beetles carrying their own brand of fungal plague into a highly susceptible tree population. In time the pandemic will pass, hopefully leaving a few resistant trees from which new generations will arise. The disaster of the Dutch Elm disease has stressed the ever-lurking presence of tree pathogens and the terrible damage they can cause to our environment.

Infectious diseases and man

The historical and economic impact of disease
Microbial diseases have attacked man and his animals since early times,

and any historical examination of man must give due emphasis to their presence. Ancient Chinese, Indian and Egyptian literature dating back several thousand years gives reliable evidence of the presence of smallpox, malaria, leprosy, cholera, gonorrhoea, and tuberculosis. In the Old Testament there is mention of gonorrhoea, leprosy, psoriasis and plague.

Major early settlements of man occurred along the banks of the world's great river systems which sustained crops and man, together with a host of disease-causing organisms. By approximately 500 B.C. four major civilizations had developed in Eurasia—the Middle East, the Yellow River Valley, the Ganges Valley and the Mediterranean coastlands. Each area possessed a wide range of endemic diseases to which the populations had achieved a degree of acquired resistance. With time, populations increased, always ensuring a reservoir of people vulnerable to infection and thus ensuring the perpetuation of the disease organisms. As these civilizations fought and traded with each other, there was gradual mixing of diseases and, by the late Middle Ages, there was a single large reservoir of infection in the Old World. When the New World was colonized by the Spanish conquistadores in the eighteenth century, the Incas and Aztecs were decimated by the arrival of hitherto unknown diseases.

Historians give little credit to the role of disease-causing microbes in the development of nations. Man with all his many forms of warfare has had less effect on the fate of nations than the typhus louse, the plague flea and the yellow-fever vector. Civilizations have retreated from the path of disease, and armies have crumbled into rabbles under the onslaught of cholera, dysentery and typhoid. In history, epidemics are blamed for defeats, but the generals take the credit for victory. The Crusaders were undoubtedly more affected by epidemics than they were by the armed might of the Saracens. The power of Napoleon in Europe was broken by disease; in particular the diseases due to bad sanitation, malnutrition and crowded conditions, e.g. pneumonia, diptheria, dysentery, typhus, and enteric fever. Retreating armies often deliberately polluted drinking water.

Even in modern times (during the struggles that led to the establishment of the Soviet Republic) the population suffered, in addition to the casualties of war, two cholera epidemics, a famine, typhus, dysentery, tuberculosis and syphilis. Between 1917 and 1923 there were 30 million cases of typhoid with 3 million deaths. However, from a historical viewpoint nothing can equal the horror of the Black Death that ravaged Europe in the fourteenth century. In the three years from 1348 to 1350 this pandemic plague killed at least a quarter of the population of Europe. This was without doubt the worst disaster ever to strike mankind. The disease organism of plague (*Yersinia pestis*) is transmitted to man by fleas carried by

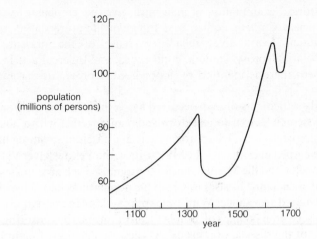

Figure 5.2 Impact on population from recurrent plagues in Europe is indicated. For more than 300 years after 1348, the plagues checked the normal rise in population. Sometimes as in the 14th and 17th centuries, they resulted in sharp reduction (from Langer, 1964).

fleas on rats. The plague struck at its deadliest in the crowded insanitary cities and towns. Paralleling the plague were often epidemics of typhus, syphilis and "English Sweat" or influenza. The mortality rate in Western and Central Europe was so high that it took nearly two centuries for the population level of 1348 to be regained (figure 5.2).

Outbreaks of epidemic plague led to unprecedented waves of crime and violence; even quiet university towns such as Oxford were the scenes of "lewd and dissolute behaviour".

The economic effect of the Black Death gripped an entire continent. The immense loss of the labour force crippled agriculture, though the scarcity of the farmer or serf greatly improved bargaining power and led to higher wages. Vast areas of land remained uncultivated, and thousands of agricultural villages were deserted.

The Black Death led Europe into a long depression and for over a century there was economic stagnation and decline. The appalling degree of death had immense psychological effects, gripping all Europe in an apocalyptic mood. The sufferings and reactions of humanity during outbreaks of plague have been vividly depicted in the writings of Boccaccio and Defoe, and by many artists including Raphael and Delacroix.

A curious by-product of the Black Death in England was the revival of English as the natural language. Since the Norman Conquest the language of educated and governing classes had been French. The Black Death swept away the upholders of the old traditions; when it receded, the national language was English.

The nature of infectious diseases

The causal microbes of infectious disease attack us from the environment in which we live, from our food and drink, and from the bodies of other people, animals and insects. However, advances in public health and sanitation, and the medical discoveries of the nineteenth and twentieth centuries have seen a great reduction in, and the near-elimination of, many infectious diseases. In particular, in the developed nations, tuberculosis, cholera, diphtheria, dysentery and typhoid have been brought under control, while in developing countries the devastation wrought by these diseases is now much reduced.

Nutrition and environment are still of paramount importance in infection. In developing countries, the common infections such as diptheria, whooping cough, measles and typhoid are respectively 100, 300, 55 and 160 times as common as in developed countries.

Although there is a tendency to believe that antibiotics and vaccines are the main reasons for disease control, there is no doubt that the improved conditions of sanitation, production of pure drinking water, disposal of sewage and, above all, the improvement in housing have also been important. The breakdown of any of these services can readily lead to outbreaks of disease. Disease-causing microbes are always somewhere in our environments. When the defences are down, the microbes, by using their ability to propagate, can rapidly build up to disease-causing levels. In former times, mankind could not understand the nature of infectious diseases, and became trapped in a world of terror, believing that disease was the result of supernatural forces. In times of severe pestilence, many sought solace in a form of religious fanaticism, others fled in blind panic, often carrying the pestilence with them, while still more sought to blame minority groups for spreading disease and subjected them to appalling victimization. With the gradual understanding of the microbial nature of many diseases, it became possible to approach the problem of infection rationally, and modern medicine now has a firm control of most microbial diseases.

Many types of microbes live in close contact with man and can be found in large numbers growing in the mouth (10^{10}) and intestine (10^{14}), or colonizing the surface of teeth and skin (10^{12}). Such microbes are existing as parasites, obtaining all of their nutritional requirements from the host. In most cases they are quite harmless, causing no apparent disturbance to the metabolism and wellbeing of man. They have achieved a truly successful form of parasitism which permits their existence, multiplication, and dispersal without unduly damaging the host organism. Only a very small proportion of the microbes associated with man give rise to serious pathological response or disease. However, when such unbalanced parasitism does occur, it can lead to destruction of tissues, poisoning of essential systems, and in many cases to death. It is this virulence or pathogenicity of certain microbes that makes them such problems to mankind.

Thus human populations and disease-promoting microorganisms live together in a state of doubtful equilibrium. In most diseases the effects are not lethal; the pathogen does not destroy the living host on which it relies for all its sustenance. Furthermore, within the populations there are variable degrees of resistance to the pathogen, not all individuals showing the same response. (The nature and diversity of man's inherent mechanisms for dealing with invading pathogens will not be considered in the present context.) Thus even in times of highly virulent diseases, e.g. the Black Death, individuals survived. It must also be noted that the enormous reproductive ability of most microbes can lead to genetic change producing new degrees of virulence. New virulent strains of a disease organism can quickly attack formerly resistant hosts.

A further point of interest concerns the global aspect of disease. In former times many diseases were restricted to specific parts of the world where populations had achieved a balanced state of pathogenicity. However, the entry into these environments of non-resistant individuals from other geographical areas can have catastrophic effects, e.g. yellow fever and Europeans; tuberculosis and Africans; and measles and Eskimos. History abounds with such examples and in our present society, with rapid inter-continental travel so prevalent, the local aspect of a disease is of less importance. Once a pathogen has entered the human body it does not really matter whether that body is in equatorial Africa or Golders Green, since the body temperature and the internal environment are the same. Thus infection may take place in one part of the world, and incubation and outward signs of the disease manifest themselves thousands of miles away. In this way diseases have lost much of their geographical limitations. Control of inter-continental disease transmission can be assisted by quarantine, whereby man or animals are held in isolation for specific periods to ensure the full development of any disease that may be incubating. The Venetians were the first to introduce the concept of quarantine during periods of plague and infestation. Since Christ had spent 40 days and nights in the wilderness, they also made the period of isolation 40 days, and the name is derived from the Italian *quarantina* meaning forty.

Most disease microbes which involve man as the true and only host have generally evolved to less pathogenic forms. Disease organisms that have not evolved to a less pathogenic form are largely those that do not require man as an essential host, e.g. those causing rabies, plague, and psittacosis have another host for their maintenance. Occasionally man will become infected by rare diseases mainly found in other animals, e.g. Lassa fever and the Marburg disease, acquired from African rodents and monkeys respectively.

Before discussing the relationship between the environment and entry of microbes into the human system, it is perhaps pertinent to examine the nature of disease transmission and also to examine briefly some of the animal and insect vectors that are responsible for infecting man with microbial diseases.

Microbial diseases may be spread in a horizontal or vertical manner. Horizontal spread of infection occurs between individuals by involving water (e.g. cholera, dysentery); air (e.g. influenza); food (e.g. typhoid, food poisoning) and by vectors of many types (e.g. malaria, typhus, plague). In contrast, vertical spread occurs by way of parents infecting offspring through sperm, ovum, placenta and milk (congenital rubella, leukemia viruses). Thus in horizontal transmission the environment of man can be of major importance in determining the extent of disease. For instance, contamination of drinking water by disease organisms can lead in a short time to rapid transmission in a population; a single infected influenza victim can infect large numbers of people in crowded stuffy atmospheres, due to the virus being transmitted in the vast aerosol sprays caused by coughing and sneezing. In contrast, venereal diseases (which must always involve close intimate contact) have a much slower pattern of infection due to the obvious physical limitations of an infected individual achieving contact with many individuals. However, a greater degree of sexual promiscuity in a population containing venereal diseases will greatly enhance disease spread. Correspondingly, the higher the moral standing of a community, the less opportunity will be available for transmission.

Man may become infected by microbial diseases from contact with other forms of life, including domesticated animals, vermin and insects. Domesticated animals can harbour diseases which can be transmitted to man and cause diseases of a moderate to deadly nature. In particular, cows can transmit brucellosis and anthrax, while dogs can transmit rabies. In the second case, the virus is transmitted by the bite of the mad dog and must be implanted directly into the human nervous system to produce the true disease.

Vermin constitute one of the most serious menaces to man's existence, since they transmit many diseases, as well as causing enormous economic losses due to biodeterioration of stored products. Rats are particularly at home in man's environment and can be found in houses, shops, factories, farms and sewers. The ability of rats to move freely between sewers and man's dwellings creates the potential for mechanically passing microbes between these environments—in particular, dangerous intestinal organisms present in faeces.

From a historical viewpoint, the ability of the rat to carry the flea which can, in turn, be the host or carrier of many diseases has had a monumental influence on man (as was discussed earlier in the context of the Black Death). That such diseases are now well controlled in most developed countries is to the credit and vigilance of our medical and sanitary services. The control of flea-borne diseases must always be by way of controlling the rat. Improved sanitation and personal and communal hygiene have all assisted. However, mankind can never relax the vigilance against vermin, and every effort should be made to exterminate them. The rat is the complete enemy of man.

The war against insects has been one of man's longest activities, and it is only in recent times that he has gained the upper hand in some aspects with the advent of powerful chemical insecticides. The physical damage to crops and materials by insects is vast, but is outwith the scope of this chapter. All insects that feed on human blood, e.g. bed bug, flea, louse, can be mechanical vectors of pathogenic microbes, and the irritation caused by the puncturing of the skin can lead to pathogenic bacteria entering the wound; chronic staphylococcal infections can occur in this way. The louse can also be the transmitter of the deadly typhus fever. Fortunately it has been eradicated from most places, but always seems to reappear in times of war and famine in cold temperate climates.

Possibly one of the greatest scourges of mankind has been the protozoal disease *malaria* which until recent years made large tracts of land unsuitable for habitation. Over two million people died annually from this disease. The vector in this case is the mosquito, which can now be controlled by insecticides and destruction of breeding centres. In temperate zones, probably the most continually dangerous insects are the carrion flies, horse flies and bluebottles that feed on human and animal faeces and other putrefying material; most pathogenic microbes that infect man are able to pass through the fly's intestinal tract unaltered. Indeed flies are the greatest carriers of summer diarrhoea and should be exterminated. Recent visitors to China have been amazed to find so few flies. This results from continued national campaigns in which all citizens are expected to show evidence of involvement.

The pathway of disease
Infectious diseases may be divided into three main groups.

(1) Microorganisms possessing unique mechanisms for attaching to and penetrating the protective body surfaces of the host.

(2) Microorganisms introduced into the host by the bite of an animal or insect vector.

(3) Microorganisms with no specific mechanisms for entering the host, which rely on some damage or breakdown to the defences of the host.

A small number of diseases do not require the entry of the disease-causing organism into the body tissues. In these examples, the organisms are present in body cavities (e.g. food-poisoning bacteria in the stomach or diphtheria bacteria in the throat) and exert their disease symptoms by excreting toxins which then enter the cells. A remarkable feature of acute infectious diseases is that they are mostly respiratory or intestinal in nature. Although not the severest of infectious diseases, they are by far the most numerous, e.g. cholera, dysentery, tuberculosis and influenza.

Skin forms the main protective layer around the vital tissues of man and serves to exclude most types of microbe present in the environment, even when they are brought into close contact. However, breaks can occur in this layer due to wounding, either accidentally or by design, and microbes can enter, causing infectious patterns. Thus in dirty unhygienic conditions there will be a greater opportunity for such chance infections to occur. Biting arthropods, such as fleas and mosquitos, can easily puncture the protective skin layer and so introduce pathogenic microbes into body tissues. Such infections can be either mechanical (when no multiplication of the microbe occurs in the vector) or biological, when the microbe can multiply in the vector, as in yellow fever and malaria. Bites of large animals may result in local infections, but these are rarely fatal, although the bite of a mad dog infected with rabies can be lethal. The spread of this disease through Western Europe now threatens Britain, and there is little doubt that in the very near future the disease will become endemic. In this disease the virus is shed in saliva and enters the human system by way of the wound.

A human lung will inhale about 10 000 microbes per day. Most are non-pathogenic bacteria and mould spores, and are easily removed from the lungs by the mucociliary defences, or are phagocytozed and are ultimately carried back to the throat to be swallowed. For a microbe to be able to initiate an infection in this part of the body, it must first attach itself to the cell, e.g. the whooping cough bacterium. Many types of disease can only become infectious if first the mucociliary system is damaged. This may occur by virus infections such as measles or influenza, or as the result of cigarette smoking or atmospheric pollution. Patients with asbestosis show an increased susceptibility to such respiratory diseases.

The intestinal tract contains a seething mass of microbes which grow

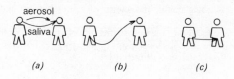

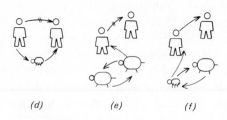

Figure 5.3 Types of transmission of infectious agents.
(*a*) Respiratory or salivary spread: not readily controllable.
(*b*) Faecal-oral spread: controllable by public health measures.
(*c*) Venereal spread: control difficult because it concerns social factors.
Zoonoses: human infection controlled by controlling vectors or by controlling animal infection.
(*d*) Vector (biting arthropods)
(*e*) Vertebrate reservoir
(*f*) Vector-vertebrate reservoir
(from Mims, 1977).

luxuriantly on the digesting food material, their multiplication being counterbalanced by the continuous movement of the bowel contents to the exterior; the more rapid the movement, e.g. during diarrhoea, the lower will be the bacterial count. If an organism is going to become infectious in the intestine, it must become attached to the epithelial layer as does the cholera bacterium. Intestinal infections will almost always arise from contaminated water or food, the pathogenic microbes passing through the stomach and entering the intestinal tract where they become attached and multiply. Fortunately many mechanisms exist within the body to ward off most of these potential infections, e.g. bacteriocins and microbial competition.

In urinary-tract infections the gonococcal bacteria owe much of their success to a special ability to attach to cell surfaces. Other types of urinary infections are at least 14 times more common in women than men due largely to the fact that the urethra which is 8 inches long in man is only 2 inches in woman and can be more easily traversed by invading microbes.

When considering pathogenic diseases, an all-important part of the infectious cycle is the exit of the disease organism from the host body, re-entering the environment and continuing the cycle of infection. The transmission of a pathogenic microbe depends largely on the manner of shedding and on its own stability in the hostile environment outwith the host. Types of transmission are shown in figure 5.3.

Many respiratory infections reach the environment in airborne particles or aerosols. A good sneeze can produce up to 20 000 particles and, if the host is infected with influenza virus, then it is not difficult to understand how rapidly infections can occur. Fortunately most respiratory pathogens are highly unstable and quickly die, though a few, such as the tubercle bacillus and the smallpox virus, are quite hardy and can cause later infections. Pathogenic organisms will be regularly shed in faeces; where diarrhoea prevails there will be an even greater opportunity of transmission and spread. Thus in environments low in hygiene there is rapid recycling of infectious-disease organisms contaminating food, water and living areas. Many infectious microbes are resistant to drying and remain in the environment for long periods. Some, such as the tetanus spore, can be highly resistant and remain viable in soils for years. Transfer of microbes via urine, faeces, food and ectoparasites has greatly decreased, at least in developed countries, but aerosol and mucosal (kissing, venereal) transfer continue.

Harnessing the microbe

Not all microbes are harmful to man and his economy. Many microbes possess useful and sometimes unique chemical activities that have been harnessed to yield products indispensable to our present way of life. Huge national and multinational corporations are centred around specific microbial processes, and the financial implications in modern society are inestimable. Only a few of the more important processes can be examined in the present context.

Microbes and fermentation
The ability of yeast microbes to convert sugar compounds into alcohol has had an immense and continuing impact on mankind. All tribes of the world (except the North American Indian and the Eskimo) discovered this process that was later to become known as *fermentation*. The production of potable alcoholic beverages is world-wide and varies only in the raw material. The Japanese prepare saké from rice, the Africans pombé from millet; in Central America a beer is produced from corn, in Mediterranean regions wines are prepared from grapes, and in Europe barley

is used in the production of various types of beers. In almost all cases the causal microbes are species of one group of yeasts, *Saccharomyces*. All these alcoholic beverages are normally quite low in alcoholic content but, with the development of the still, it became possible to prepare high-alcoholic drinks such as whisky, rum, brandy, gin and vodka. Distillation apparently developed during the ancient Egyptian civilization.

Pasteur, in the middle 1800s, finally established that the chemical process of alcohol formation or fermentation resulted from yeast action. Pasteur's "germ theory of fermentation" came as a bolt from the blue, and gravely disturbed the scientific communities who hotly disputed his findings and ferociously clung to their own theories of chemical interaction and spontaneous generation. By careful painstaking experimentation Pasteur demonstrated the causal relationship and showed how other microbial contaminants could alter the chemical balance of wine and make it "sour". He found that if wine was heated to 55°C the microbes causing souring were destroyed, but the taste of the wine remained unaltered. This process of *pasteurization* is now widely practised in other forms of perishable liquids, including milk. Unlike sterilization, which aims to destroy all living organisms in a substance (e.g. tinned foods) so that it keeps indefinitely, pasteurization aims only at partial sterilization to prolong the keeping properties of perishable substances. The advantage of pasteurization is that it has very little effect on the taste and properties of food and liquids.

The social and economic influences of alcoholic beverages have reverberated throughout the ages. Over-indulgence and misuse abound, and the Bible bears witness to such practices. The narcotizing and stupefying influences of alcohol always have been a handmaiden of poverty, serving as an escape from reality. For many years nations have attempted to legislate on its manufacture and distribution—an example being prohibition in the United States. In most Western countries the problem of alcohol addiction is real, and the loss of production due to absenteeism is substantial. Alcoholism in the young is now a serious social problem and in many countries it is probably a greater problem than drug addiction. Huge industries have developed around the distillation, brewing and wine-making communities, and they represent a major source of revenue for national economies. The sudden arrival of a microbial epidemic of grapes (witness the downy mildew in Europe), rice or barley, or microbial infiltration into the production systems can cause immense economic disturbances.

It is surprising to note that only a part of microbial alcohol production is used for beverages—most is used in industrial processes. As a solvent, alcohol is used in the manufacture of dyes, drugs, soaps, plastics, and resins, and as a raw material in the synthesis of ethers, esters and acetic acid.

Microbial fermentation using other yeasts and bacteria can produce many other alcoholic brews, including propanol, butanol and vinegar. New fermentation processes are being examined and, in the present energy crisis, there is increasing interest in converting cellulose by microbial fermentation into simpler and more usable compounds. Fermentation will continue to serve mankind in the future and, as world oil production decreases, so will fermentation practices increase.

Microbes and chemotherapy

The science of controlling microbial diseases by specific chemicals (chemotherapy) owes much to the spectacular discovery by Paul Ehrlich in 1910 that a drug called salvarsan could selectively kill the syphilis bacterium. Yet another success of the chemotherapists were the sulphonamides or sulfa drugs which are active against the bacteria causing pneumonia and also puerperal (or childbed) fever. Chemotherapy languished for many years and, although most major pharmaceutical companies invested heavily in the search for more selective compounds, successes were few; although many synthetic drugs were capable of killing the microbes causing disease, they had serious and often lethal effects on the host.

The discovery by Alexander Fleming that a mould called *Penicillium* could produce a compound able selectively to inactivate a wide range of bacteria without unduly influencing the host, set in motion scientific studies that have profoundly altered the relationships of man to the controlling influence of bacterial diseases. From these studies have emerged penicillin, streptomycin, aureomycin, chloramphenicol and the tetracyclines. As a result of using these antibiotics, bacterial diseases, given correct medical supervision, are now very much under control. Tuberculosis, pneumonia, venereal diseases, cholera and leprosy, to mention only a few, no longer dominate man and, at least in the developed parts of the world, have been relegated to minor diseases. Griseofulvin, an antibiotic active against fungal diseases, has brought great relief to those infected with debilitating fungal skin diseases such as ringworm. Ringworm of the body and scalp has long plagued mankind and is predominantly a disease of the young, particularly in crowded unhygienic conditions.

Although the chemotherapeutic control of many diseases has been a salvation, relieving man from the miseries of disease, it is in part responsible for the present world population crisis. Apart from direct saving of lives, the prolongation of life, particularly in early childhood, has vastly extended reproduction potential and is contributing to the present explosive world population.

A disturbing and disquieting observation has been the gradual evolution of drug resistance in many microbes. The indiscriminate use of antibiotics for minor bacterial infections can no longer be condoned. The possibility of an acquired resistance being transmitted to another type of microbe is now real. Gonorrhoea resistant to treatment with penicillin is now present in 19 countries, including Britain, and it is feared that the resistance factor may be transmitted to close bacterial relatives which cause meningitis.

The discovery of the antibiotics meant much to mankind but, because of such phenomena as drug resistance, the search for new antibiotics must always continue. Antibiotics are used to treat established diseases. In contrast, vaccines are used to prevent disease and, when used on a large scale, have become an important antimicrobial power in our community. A *vaccine* is a material, originating from a microorganism or other parasite, that induces an immunologically mediated resistance to disease. Vaccines are prepared from the causal organisms of the disease and have been effective against smallpox, poliomyelitis and yellow fever; less effective against typhoid and cholera; and completely unsuccessful towards malaria, syphilis and gonorrhoea.

Microbes as food

It is now widely accepted that the major problem facing the world—and particularly the developing countries—is the explosive rate of population increase. In consequence there is a massive problem facing conventional agriculture to supply sufficient food and, in particular, protein to satisfy these demands. The FAO have already predicted a widening of the protein gap between developed and developing countries. Much effort is now centred on producing protein from unconventional sources including:

(a) Mechanical extraction of proteins from coconuts, soybeans, groundnuts and many types of leaves
(b) Protein production from non-domesticated herbivores
(c) The conversion of animal products such as wool and feathers into food
(d) The production of microbial protein for food purposes, in particular the feeding of man's domesticated animals.

Man has long recognized the nutritional value of some microbes, namely the fungi. Although a few mushrooms are poisonous, the great majority are edible and have a good protein content. However, scepticism and prejudice have for centuries dominated man's outlook on mushrooms. Mushrooms are eaten freely in many countries, while in others they are avoided and neglected. The hieroglyphics of the Egyptians record legends indicating a belief that certain mushrooms were the plants of immortality, and the Pharaohs, respecting the delicious flavour, decreed that they were forbidden to commoners.

Man first produced the white mushroom (*Agaricus bisporus*) commercially in and around the caves of Paris. For many years only the French possessed the art to grow this culinary treasure, and it was enjoyed only by the rich. Gradually the cultivation of mushrooms spread to Britain and America, and the art of mushroom cultivation gave way to a sophisticated technology. Traditionally, mushroom cultivation has been centred in Europe and America, but in recent times cultivation has spread to Taiwan, Australia, New Zealand, Korea, China, South America, India, and some Mediterranean countries. World production is now estimated to be in excess of 300 000 metric tons per annum, with an annual increase of over 10%.

Several other types of fungi have been developed commercially. The Japanese excel in growing the Shi-ta-ké (*Lentinus edodes*) mushroom on tree logs; throughout Asia the Padi Straw (*Volvariella volvaceae*) mushroom is grown by simple techniques on piles of rice straw, and it supplies some essential protein in the diet; the truffle (*Tuber melanogaster*) grows naturally below ground on the roots of oak trees in many parts of France and Italy.

Thus man is no stranger to the use of certain fungi for food. Their cultivation will increase, but the cultivation procedures are complicated and expensive, and they must be considered largely as a source of flavour rather than for protein. In recent years there has been an explosive interest in using other forms of microbes for food production, in particular for the feeding of domesticated animals. Many major companies in Europe and the United States are actively involved in those studies, and already many worthwhile products are commercially available. Protein quality and quantity are the goals of these studies.

The advantages of producing food from microbes are as follows:

(i) Microorganisms have the ability to grow at remarkably rapid rates under optimum conditions; some microbes can double their biomass every $\frac{1}{2}$-1 hour.
(ii) Microbes can be more easily modified genetically than can plants and animals. Thus it becomes possible in large-scale screening programming to select microbes with higher growth rates, better amino-acid content, etc.
(iii) Microbes can have relatively high protein content.
(iv) Microbes can be grown in small continuous processes using relatively small land area and can also be independent of climate.
(v) Microbes can grow on a wide range of raw materials, in particular many waste products, and can also utilize plant cellulose.
(vi) The nutritional value of the protein is good.

Current investigations permit a broad classification into three main food sources for microbes:

(a) Materials with a high commercial value as energy sources or derivatives of such materials, e.g. gas-oil, methanol, ethanol, methane, n-alkanes. The microbes involved are mostly bacteria and yeasts, and several successful processes are now in operation. Most oil companies have interests in this field and recently I.C.I. entered the scene and will shortly be producing 50 000 tons of bacterial protein (Pruteen) per annum from methanol, to feed animals and poultry. The wisdom of using such high-energy-potential compounds for food production has been questioned by many scientists.

(b) The second main group of materials may be considered as waste and should be recycled back into the ecosystem, e.g. bagasse, straw, citrus waste, whey, olive and date waste, molasses, animal manure and sewage. The amount of such wastes is very high and can contribute in many areas to a significant level of pollution. Thus the development of methods to utilize these materials serves two functions: reduction in pollution, and creation of edible protein. The microbes mostly used in this context are yeasts and filamentous fungi. The products of these microbial interactions could well become competitive with traditional animal feedstuffs.

(c) The final category of materials include those that can be derived from plants, and hence are a renewable source, e.g. starch, sugars and cellulose. Here lies the greatest potential for protein production; perhaps in most tropical areas plant crops will be produced simply to be utilized as substrates for microbial fermentation. Indeed it has even been considered that the giant petrochemical industries could obtain all their basic substrate from microbial breakdown of complex plant materials. The energy barons of the future may well be the countries with good crop production potential, such as Brazil. Filamentous fungi, with their great ability to break down cellulose, have been particularly used in this field. RHM Bakeries, utilizing starch, already produce a fungal product which has great potential as a food for man.

Many everyday goods that we consume are the result of microbial action. Milk is an ideal mixture for the growth of many microbes and, whereas some may be the causative organisms of disease in man (e.g. brucellosis), others can convert milk into yoghurts, cheeses, butter and buttermilk. Cheeses have long been part of man's diet, and many of the most famous, such as Roquefort and Camembert, owe their unique flavour to microbes endemic to particular parts of France. Now that the microbial involvement has been recognized, however, it is possible to make most cheeses anywhere in the world. The leavening properties of yeast have been long appreciated, and now form the basis of extensive baking industries.

Finally, mention must be made of the microbial foods made in the Orient, from beans and other plant products. Such foods are usually the end result of mixed microbial reactions, including yeasts, bacteria and fungi, and have been part of the diet for hundreds of years. Not only are these products nutritionally good but they are often highly flavoured and can be added to bland diets.

FURTHER READING

A. Books of general interest

Birch, G. G., Parker, K. J., and Worgan, J. T., eds. (1976), *Food from Waste*, Applied Science Publishers.

Carpenter, P. L. (1972), *Microbiology*, 3rd ed., W. B. Saunders Co.

Dickinson, C. H. and Lucas, J. A. (1977), *Plant Pathology and Plant Pathogens*, Blackwell Scientific Publishers.

De Kruif, P. (1950), *Microbe Hunters*, Pocket Books Inc., New York.

Drew, J. (1960), *Man, Microbes and Malady*, Pelican.

Lovelock, D. W. and Gilbert, R. J. eds. (1975), *Microbial Aspects of the Deterioration of Materials*, Academic Press.

Mims, C. A. (1977), *The Pathogenesis of Infectious Disease*, Academic Press.

Porkiss, B. E. *Biotechnology of Industrial Water Conservation*, M. & B. Monographs CE/8 London, Mills & Boon Ltd.

Postgate, J. (1975), *Microbes and Man*, Pelican.

Walters, A. H. and Elphick, J. S. eds. *Biodeterioration of Materials*, Vol. 1 and 2 (1971). Applied Science Publishers.

Zinsser, H. L. (1934), *Rats, Lice and History*, Bantam Books.

Reviews of specific interest

Hacking, A. and Harrison, J. (1976), "Mycotoxins in Animal Feeds," in *Microbiology in Agriculture, Fisheries and Food*, edited by F. A. Skinner & J. G. Carr, Academic Press, pp. 243–250.

Jarvis, B. (1976), "Mycotoxins in Food," in *Microbiology in Agriculture, Fisheries and Food*, edited by F. A. Skinner & J. G. Carr, Academic Press, pp. 251–267.

Langer, W. L. (1964), "The Black Death," *Scientific American*.

Smith, J. E. (1976), "Microbial Spoilage of Engineering Materials," *Tribology International*, pp. 225–230.

Smith, J. E. (1969 & 1972), "Commercial Mushroom Production," 1 & 2, *Process Biochemistry*.

CHAPTER SIX

CONSERVATION OF THE BIOLOGICAL ENVIRONMENT

J. N. R. Jeffers

Introduction

Conservation of the biological environment has emerged as one of today's principal political, social, and economic issues, but it is only relatively recently that, apart from a small number of far-sighted individuals, much attention has been given to what has been called "the wise principle of coexistence between man and nature, even if it has to be a modified kind of nature". Even today, wildlife and environmental conservation are dependent upon a well-developed economy, and are only advocated with enthusiasm by people who already have enough to eat, sufficient protection against heat or cold, and an adequate supply of material comforts. Indeed, the less developed countries of the world are often critical of those who urge conservation of the biological environment on peoples who have not yet reached reasonable standards of living. Sometimes associated with demands for the conservation of the biological environment is the development of a mode of thought which is characterized as "ecological", and which denotes a generalized popular sense of a philosophy, a way of life of quasi-religious dogma, that asserts that what is "natural" is good and that what is not "natural" is bad. The conjunction between the conservation of the biological environment and "ecology" in this sense is, however, false. The original meaning of ecology is of a scientific discipline which studies the ways in which plants and animals are distributed, associated in communities, and related to the physical and chemical factors of their environment. The aim of this study is to guide the ways in which people modify their environment so that they do not unwittingly damage elements of it on which they depend, or which they cherish, and to enable them to

122

make choices which ensure that they live in the best attainable world.

In this chapter, we shall first examine the ecological basis of our understanding of the biological environment. While it is true that much popular support for conservation is expressed in terms of individual organisms, and particularly the more attractive plants and animals, it is the assemblage of these organisms into communities which will necessarily occupy most of our attention. Within these communities, it will be possible to identify food chains, and the roles of primary and secondary producers and consumers. These roles increase our understanding of the ways in which nutrients and other substances pass through the communities, and help to identify where various substances may be stored. The organization of communities into biomes also helps to focus our attention on the larger-scale changes related to climate and to man's use of his environment. Our particular concern will be with the succession from one ecological system to another, frequently in response to deliberate perturbations brought about through management by man.

Next, we shall consider some of the strategies which have been used for conservation, both nationally and internationally. While there is considerable variation from country to country in the extent to which attempts have been made to conserve the biological environment, progress in this important field has been more advanced in Britain than anywhere in the world. We shall therefore concentrate on the sequences of events that have taken place in Britain, and the approaches that have been made towards effective conservation. Similarly, we shall concentrate on a relatively small number of the international initiatives in biological conservation, and particularly on the *International Biological Programme* and on the *Man and the Biosphere Programme* of the United Nations Educational, Scientific and Cultural Organization (UNESCO).

Having reviewed some of the approaches taken to conservation, it will be appropriate to consider in more detail the genetic basis for conservation. This review will quickly lead us to a study of man's impact upon the biological environment in urban and rural areas. Man's perception of the basic problems greatly affects the extent of the impact that he will have upon the environment, and will ultimately depend on the extent to which he is successful in predicting the results of his manipulations of marine, freshwater and terrestrial systems.

Ecology and ecosystems

Communities

Much of the "popular" interest in wildlife conservation is focused on individual species of plants and animals. Usually these are the more attractive plants and animals, although "attractiveness" is enhanced by rarity and becomes hard to define. Nevertheless, the appeal of conservation is more likely to be greater for flowering plants than for mosses, liverworts or inconspicuous soil fungi, and greater for large mammals and birds than for worms or biting insects. Rarity, too, can be misleading if it refers only to a limited locality. Much effort can be devoted to conserving a plant or animal at the extremes of its geographical range for the sake of recording it as "present" in a particular country or region, despite short-term climatic or other changes.

For these, and other, reasons it is important to consider assemblages of plant and animal species into communities, and to study the interrelations of these plants and animals with each other and with their typical environment—a statement which is itself a definition of ecology. The interaction of living organisms with their environment is very much a two-way process, in that organisms affect their environment and are in turn affected by their physical surroundings—a process which is known to engineers as *feed-back*. For this reason, a British botanist, Sir Arthur Tansley, suggested in 1935 that the term *ecosystem* should be used to describe "... not only the organism complex but also the whole complex of physical factors forming what we call the environment". Modern ideas about the conservation of the biological environment depend upon the concept that it is useless to conserve individuals of a particular organism unless you also conserve the ecosystem which enables that organism to survive.

The concept of the interaction of living organisms with the physical and chemical factors of their habitat has proved to be one of the most important ideas of science, and the ecosystem has become the basic unit of ecological studies. Various ways of classifying and studying ecosystems have been proposed, depending upon the scale of investigation and the primary focus of attention. Essentially, however, the role of any organism will depend on its place in the ecosystem, and the conservation of organisms, communities of plants and animals, or whole ecosystems, depends upon our understanding of the interrelationships. Any attempts to interfere with these relationships, for the purpose of conservation, production or harvesting, without adequate understanding, run the risk of upsetting the whole balance of a delicate network of dependencies, as has been illustrated many times in man's short history.

A modern example of the kind of problem that can be posed by lack of understanding of ecological relationships is given by the present condition of the Sahel Zone in West Africa. Intensified animal husbandry has been made possible by various successful measures in combating the tsetse fly— seen as a desirable improvement of the economic and social welfare of the people. The sudden boost given to the nomadic economy by these measures, however, has led to over-grazing, to population growth, and to concentration of cattle herds around the artesian wells, so that the supply of water for humans, animals, and vegetation was disrupted. The resulting damage to vegetation has led to further unfavourable effects on the region, where rainfall is already uncertain, with consequent loss of animals, crops, and human lives.

The construction of the Aswan Dam in Egypt, as part of a great land irrigation project, is another example of a plan which lacked sufficient ecological consideration. A delicately-balanced cross-linkage of inter-actions resulted in a rapid spread of the water hyacinth, which acted as a breeding ground for the snails which carried bilharzia and caused excessive evaporation of water from the reservoir. The water from the storage lake was so low in nutrients and silt that artificial fertilizers had to be used in the Nile Valley, while the increased flow of water eroded the river bank. What started as a scheme for the improvement of agriculture and standards of living has resulted in a new series of ecological problems which have not yet been solved.

Trophic levels

One important way of studying ecosystems is by grouping their component organisms according to their methods of feeding or obtaining energy. Each group of organisms represents a *trophic level*, and is defined by the method of nutrition adopted by all members of the group. The green plants which obtain their energy from the sun are the *primary producers* and form the first of these trophic levels. The herbivorous animals, which range from minute invertebrates to large mammals, feed on these living plants and are described as being *primary consumers*; they occupy the second trophic level. Other animals prey upon the herbivores and form a third trophic level and, in turn, these carnivores are preyed upon by top or predatory carnivores to produce a fourth trophic level. The decomposers and detritus-feeders form yet another trophic level which accepts the residues from the other levels and turns them back into nutrients to be used by the primary producers, together with the sun's energy, to create new energy for the whole ecological system. Each box in figure 6.1 represents a trophic level, and the arrangement and relative sizes of the boxes show the trophic structure of a hypothetical ecosystem. All ecosystems possess a characteristic trophic structure, and may be studied by the investigation of that structure. In the example of figure 6.1, the flow of energy through the ecosystem is illustrated by the solid lines and arrows. All of the trophic levels lose energy from the system into the atmosphere through respiration.

Some organisms do not fit conveniently into a single trophic level. Bears,

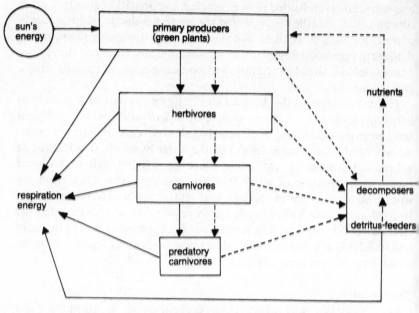

Figure 6.1 Energy and nutrient flow through trophic levels

for example, will often eat small animals, fish, roots or berries, according to their relative abundance at any particular time. Assignment of such organisms to a particular trophic level will therefore depend upon the source of food which is regarded as important in a given situation. Some organisms, including man, may feed on two or more trophic levels at the same time. In most systems, the only external source of energy is the sun, but, for many sub-systems, energy enters in the form of live or dead organisms, or in the form of decomposed organisms from another system. The energy is used by organisms for synthesizing new compounds for growth and reproduction, and also to maintain the cells in their bodies, for movement, and to maintain body temperatures. The energy for these processes can be made available through breakdown of organic molecules in respiration, but not all the energy released in this way is utilized by the organisms, so that a proportion is always lost and dissipated as heat. As a result, there is a constant flow of energy through the ecosystem from primary producers to carnivores and decomposers, and a constant loss of energy to the atmosphere as a result of respiration.

The energy fixed or absorbed by organisms is of fundamental importance in defining the amounts of energy available for passing through the ecosystem. Ecosystem productivity is therefore much studied

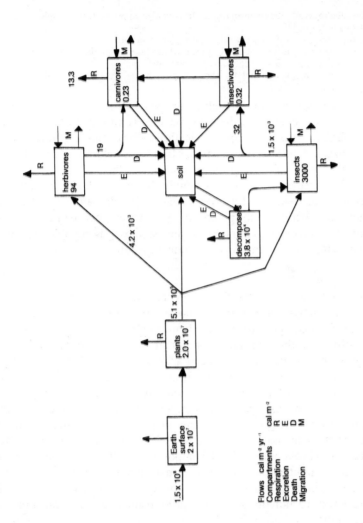

Figure 6.2 Energy flow in lowland tundra.

by ecologists, and involves the measurement of the flow of energy between trophic levels. This study is probably the most popular framework for comparing the productivity of ecosystems, and seeks to answer such questions as:

(a) How much energy enters a given trophic level in a given period (for example, a year, or a growing season)?
(b) What proportion of this energy is lost through respiration?
(c) How much energy is lost to other trophic levels by grazing, predation, or parasitism?
(d) How much energy is lost to decomposer organisms?
(e) How much energy accumulates through growth and reproduction of the living organisms?

From such information, the efficiency with which the organisms in a particular trophic level use the available energy can be calculated. Such knowledge can have important implications. For example, studies on the larger grazers of the East African plains (mainly antelope and zebra) show that these animals convert grass to meat more efficiently than do cattle kept by local tribes. As another example, figure 6.2 shows the flows of energy in a lowland tundra ecosystem in Canada, and identifies the transfers of energy from plants to herbivores, carnivores, insects, insectivores, and decomposers. From such studies, it is possible to predict the likely effects of various ways of managing habitats.

Nutrient cycles
It is not only, however, the flow of energy to the trophic levels of an ecosystem which is of importance in the study of ecology. An almost equal interest is focused upon the flow of chemical elements, either as nutrients or as pollutants, through ecological systems as indicated by the broken lines in figure 6.1. Primary producers take up mineral elements, such as nitrogen and phosphorus, in the form of soluble mineral salts from surrounding soil or water. Herbivores and carnivores obtain nitrogen and phosphorus mainly as organic compounds in their food, although cattle and humans may require supplements of raw materials such as salt or copper. The dead organisms, together with their waste and excretary products, are broken down by decomposers (chiefly bacteria and fungi) which release mineral nutrients in a form available for re-use by primary producers. Similarly, there is a cycling of carbon released into the atmosphere as carbon dioxide as a result of respiration of plants, animals and decomposers. This carbon dioxide is taken up by green plants during photosynthesis, and the carbon passes on to animals when they eat these plants. There are similar cycles for nitrogen and oxygen, and indeed for any chemical elements which we need to consider. Figure 6.3 shows the flow of nitrogen through the same tundra lowland ecosystem as that given for the flow of energy in figure 6.2. In recent times, it has become

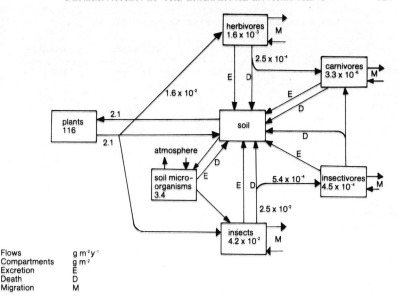

Figure 6.3 Nitrogen flow in lowland tundra.

increasingly important to study the flow of pollutant elements and compounds through ecological systems. Many of these pollutants are stored and concentrated in the tissues of predatory animals, having been acquired from other trophic levels. The mechanisms by which these substances pass through the trophic levels are therefore of particular significance. For example, many birds of prey such as kestrels, sparrowhawks, and barn owls are capable of storing considerable quantities of toxic substances such as dieldrin in their tissues by processes which are linked through photoperiodic responses.

Biomes
Important though the flow of energy, nutrient and other substances through the trophic levels of ecological systems may be, the studies of ecologists are not confined to these topics. Particular interest has always been expressed in the ways in which organisms combine in communities which are characteristic of a particular time and place, and which reflect past and present land use. The ways in which these communities are formed are complex, and are much affected by natural and man-made changes. For example, changes in climate may modify the associations of plant and animal communities, leading to a change from forest to savanna or grassland, and hence changing the whole pattern of possible land uses. Nevertheless, the plant communities and associations over the

Table 6.1 Area and productivity of the principal world biomes

Biome	Area $10^6 km^2$	Average net primary production $10^4 J m^{-2} yr^{-1}$	World net primary production $10^{19} J yr^{-1}$
Desert, rock and ice	24	6	0.1
Desert scrub	18	130	2.4
Tundra and alpine	8	270	2.1
Lake and stream	2	950	1.9
Temperate grassland	9	950	8.5
Woodland	7	1130	7.9
Cultivated land	14	1230	17.2
Tropical grassland	15	1320	19.8
Boreal forest	12	1510	18.1
Temperate deciduous woodland	18	2480	44.6
Tropical forest	20	3780	75.6
Swamp and marsh	2	3780	7.6
Open ocean	332	242	80.3
Continental shelves	27	662	17.9
Estuaries	2	3780	7.6

world as a whole can be broadly classified into what are known as *biomes*. These biomes embrace the major vegetation types of the world and, within broad limitations, have a characteristic productivity. Table 6.1 shows the approximate area, average and total primary production for the major biomes of the world.

Succession

The species composition of all biological communities varies in space and time. This variation is partly a function of the interactions between the physical and chemical processes of ecosystems, but also depends on the scale at which the community is examined. The relationship between an organism and its environment that characterizes a site in time and space becomes part of a larger mosaic when sites are combined into a wider community, and these wider communities into biomes. Nevertheless, despite the marked change in communities which can be observed from site to site, equally marked changes can be shown to occur with time, even on the same site.

This phenomenon of change in communities is called *succession*. Classical concepts of ecological succession make two key assumptions:

(1) The replacement of one species by another during succession occurs because organisms modify their environment, making conditions less favourable for their own survival and leading to replacement by other organisms.
(2) A community of organisms finally appears which is self-perpetuating, having achieved a balance of the physical and biological factors, and this final community is called a *climax*.

This view of ecological succession is based largely on changes which are thought to have occurred over centuries in natural or semi-natural communities, or in communities which have been allowed to revert to nature after some form of cultivation. It is now suggested, however, that, when ecosystems are perturbed in some way, either by man or by a natural catastrophe such as a landslide, the succession may never develop to a "climax".

In the classical theory, each successive set of species which occupies a site gradually makes the environment less favourable for its own persistence and more favourable for successor species to invade and take over. The changes due to the earlier colonizers may either increase or decrease the rates of replacement and growth of later species. These appear later, either because they arrived later or because their germination was inhibited and their growth suppressed. In contrast, in truncated successions, the early occupants, rather than preparing the way for other species, slow down the invasion of other species, either by taking up all the available space or by competition for available food, etc. The succession is then truncated at a stage at which it is generally regarded as

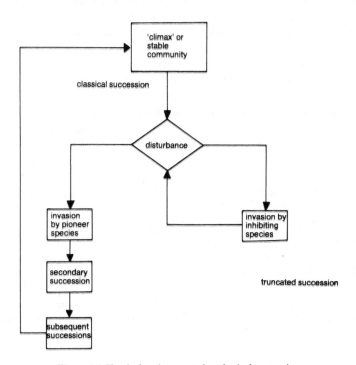

Figure 6.4 Classical and truncated ecological succession.

being composed of non-climax species. New species may only be able to enter the community when the inhibiting species are damaged or killed. If there is a subsequent perturbation, the new succession may well follow a different course and avoid a repetition of the truncation (figure 6.4).

In many communities, major disturbances occur frequently enough for the succession to be cut short, and started all over again. In Britain prehistoric man lit fires to drive out game, and cut vegetation to clear land for agriculture. Disturbances not associated with man are natural fires, landslides, severe storms and some forms of intensive insect grazing.

Within the past several thousand years, for example, much of the forest of North America has been damaged or destroyed by fire, at least once every few hundred years—a time interval which is within the lifespan of the dominant conifers. These disturbances are so widespread as to suggest that, even before man's interference became common, "climax" vegetation communities were only rarely reached.

Stability

In recovery from a perturbation it is the maintenance of the species composition which is of primary importance. However, the stability of any community depends upon three factors: namely, the frequency, the area, and the intensity of perturbations. In order to judge the stability of a community, it is necessary to decide for how long and over what area the species composition must persist for a given intensity of perturbation. The changes following a perturbation may take a variety of forms and may not return to the relatively stable assemblage of species found in the undisturbed areas. The pattern of succession, in terms of the organisms present and their relative abundance at various times after a disturbance, is influenced by the size of the area perturbed and the severity of the perturbation. The availability of an organism, either as a seed or by vegetative propagation, is increasingly dependent on highly specific biological characteristics as both the area and severity of the perturbation increase. Organisms differ widely in their ability to persist as seeds, in their ability to become established or persist in exposed and unfavourable locations, and in their competitive ability to obtain water and nutrients.

Ecologists now have a wide variety of ways of measuring the dynamic change of ecosystems, and these techniques are necessary in order to predict the likely effects of disturbances, and particularly those made by man in his attempt to improve or maintain his quality of life. In the past, we have not usually been successful in predicting the ecological consequences of changes which appear to be desirable from an economic or social point of view. As a result, the biological environment has suffered from many unexpected or unintended consequences. Even today, policies for agriculture or forestry, and for the harvesting of marine organisms, have damaging effects upon the biological environment, and upon organisms which we value for their aesthetic, amenity or conservation interest. It is even possible that measures taken to conserve a particular species, habitat, or ecosystem, can have precisely the opposite result if we do not fully understand the complex interactions of the ecological system and seek to perturb successional changes in an undesirable way. As has been indicated, the only sure remedy is a careful study of the fundamental behaviour and properties of the underlying ecological systems. Different kinds of ecological systems vary in their resilience under the stress of physical and chemical disturbance, and in their resistance to invasion by new species of plants and animals. While the checks and balances of ecological systems involving large numbers of species operating at different trophic levels are thought to give stability to climate, soil and other vital components of the environment, it requires relatively little disturbance to upset the delicate balance which is maintained by a community of organisms in harmony with its physical and chemical environment.

National measures

The conservation of the biological environment may be regarded as a national problem, transcending the boundaries of individual ownership and of traditional professional interests. We may, therefore, usefully review the kinds of measures that may be taken nationally to ensure effective wildlife and environmental conservation.

Research
The strategy for any national measures for the conservation of the biological environment must take into account several factors. First, it will be necessary to do research on the basic ecology of the habitats, communities, and organisms which are to be conserved. As has been stressed, the study of ecology is complex, and frequently difficult, because of the many interactions between organisms within an ecological system, the delicate balance of natural processes, and the way in which this balance affects living organisms and their environment. The processes of nutrition, regeneration and succession frequently require methods of investigation which take many years. The dynamics of populations of organisms, and the interrelationships between predators and prey, and between diseases and their hosts, may also need to be traced and developed over many years. The pathways of nutrients and pollutants require careful investigation over a wide variety of sites and time scales to

determine the extent to which the more harmful effects can be avoided or modified.

Nature reserves

Second, it will be necessary to establish nature reserves and to identify sites of particular scientific interest. The main assemblages of plants and animals in the country may be well known, so that it is possible to classify them and to see how they would compare with communities in other countries. By surveying the fragments of countryside that are essentially "natural", even though altered by man's activities, it may be possible to find remaining examples of relatively unmodified communities. Where the assemblages are not known or understood, it will usually be necessary to undertake extensive surveys to discover the range of communities, and the extent to which these communities are related to such factors as soils, topography, human population densities, and climate. The results of these surveys will form the basis of a list of proposed national nature reserves. Some of the proposed areas may contain impressive geological features (including gorges, landslips, and fine rock exposures) which may also require protection. Some areas will be chosen because they are unusual and contain communities of plants and animals rarely, if ever, found elsewhere. Other areas will be chosen as the best examples of typical kinds of country, for example, in Britain, of chalk grasslands, moorlands or oakwoods. Although such areas may be fairly common in one country, they may not occur elsewhere in a continent, or in the world, so that that particular country has a special international responsibility for ensuring that examples are protected. The complete series of reserves will aim to represent samples of all the more important biological environments, and to contain most of the kinds of wild plants and animals to be found in the country.

Nature reserves are sometimes purchased for the nation but, alternatively, they may be leased or established by agreements under which the owner and any tenant retain their rights, and undertake to manage the land in accordance with the advice from some designated authority. Reserves may be made available to scientists in universities and research institutes as places at which they can undertake ecological research and training, and they may be made available to the public as places where wildlife can be seen. Obviously, it is not always possible to let anyone visit such reserves at any time of the year. Some kinds of birds, for example, have to be protected from disturbance, even by sympathetic observers and scientists, and, when reserves are established under agreement, the landowners may have good reason for restricting access by issuing permits to only a limited range of people. National reserves will usually be complemented by reserves owned or leased by local natural-history societies or local conservation trusts, and will again be supplemented by sites which are known to be of particular scientific interest, notified to planning authorities and to owners. By designating a sufficient series of reserves, sites of special interest, and other designated areas, it will often be possible to safeguard a wide range of habitats, communities and organisms.

Land use

A third aim in the conservation of the biological environment will be to encourage land users in the rural environment to permit as much conservation of the better habitats as is compatible with their primary uses. There is a complete spectrum from land of national or international scientific value, managed primarily for conservation, to urban environments which support virtually no wildlife at all, and where management for conservation may be irrelevant. This situation is shown diagrammatically in figure 6.5. At the top of the triangle is the very small area of nature reserves, managed primarily for biological conservation. Immediately below is the area representing the sites of special scientific interest in which the primary land use should be compatible with the maintenance of their scientific interest. Most land, including nearly all agricultural and forest land, is in the third and largest category. At one end of this category are the good wildlife habitats which are managed primarily for agriculture, forestry or recreation, where biological conservation will be integrated as far as possible. At the opposite end of the category are the intensively-managed crop lands where wildlife will be largely confined to the soil and where management for biological conservation, other than for soil organisms necessary for agricultural or forest productivity, is not required. Thus, the requirement for management for biological conservation increases from the base upwards, but only becomes primary on a small percentage of the total land surface. The success of operations carried out on land specially scheduled for nature conservation and scientific purposes depends to a large extent on the success with which biological conservation is integrated with other land uses in the wider countryside, because many of the populations in the scheduled sites depend on replenishment from similar habitats outside them. The wildlife habitats in the wider countryside are also important in their own right, because they support most of the individuals of the commoner species of organisms.

It follows that any national measures for conservation must be directly concerned with the management of the rural environment as a whole. The ecological changes due to modern methods of forestry and agriculture are unusually rapid compared to those due to natural causes. Even so, they are not often detected by casual observers until the effects have been too

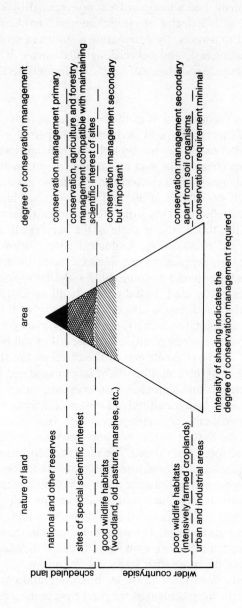

Figure 6.5 Conservation management strategy.

Table 6.2 Conservation value of various habitats

Category 1 *Most important*	Category 2 *Moderate importance*	Category 3 *Little importance*
Old woodland	Deciduous plantations Mature conifer plantations Recently planted conifer plantations Copses, corner plantations, etc. Hedges	Pole-stage conifer plantations
Lowland heaths	Moorland and rough grazing	Derelict land in towns
High mountain tops		
Unpolluted and untreated rivers, lakes, canals, permanent dykes	Gravel and clay pits Farm ponds	Polluted water of all kinds
Large marshes and bogs	Small marshes and bogs	
Permanent pastures and meadows untreated with fertilizers or herbicides	Road and railway verges Disused quarries	Grass leys Improved pasture Playing fields Airports
	Large gardens Golf courses Arable land with many weeds Neglected orchards	Small gardens Allotments Arable land with few weeds Horticultural crops, commercial orchards
Coastal habitats (cliffs, dunes, salt marsh, etc.)		Industrial and urban land

severe to be corrected. There have been, for example, extensive declines in the populations of many predatory birds, including peregrine falcons, barn owls, kestrels, and sparrowhawks, because of the use of pesticides. It was only when these effects had been recognized that it was possible to modify the use of the pesticides, and to replace them by substances which, at least at the present time, have not been shown to have such harmful effects. Similarly, the greater intensification of farming practices has resulted in larger field sizes, with the resultant loss of hedgerows and small copses. Such intensification results in severe loss or impoverishment of the habitats for plants and animals where it occurs. Table 6.2 shows the conservation value of various habitats, with a gradual reduction in the value of land which is increasingly specialized in terms of its use or, alternatively, polluted.

Advice
Inevitably, national measures for biological conservation will include advice about the ecology of plants and animals—especially, in these

days, when many open spaces are heavily used for recreation. Important aspects of this advice will be of relevance to the planning departments of local and central government. Planning (in the rural environment especially) must be based on a deep understanding of the natural patterns and processes of ecology. It must take into account the choices open to the community and the consequences of these choices. Above all, it must make clear the conflicts between policies of the many agencies that have views on, or responsibility for, the rural environment. Increasingly, too, private landowners seek advice on the management of their land in ways which are consistent with the conservation of the biological environment.

Emergencies

Finally, any national measures will inevitably be much concerned with emergencies and sudden threats to organisms and to plant and animal communities. Today, spilled oil from tankers remains a constant hazard to wildlife. The *Torrey Canyon* disaster in 1967 was a dramatic example which led to standard procedures being set up by the British Government to protect plants and animals from oil pollution in the sea. The more recent disasters of the *Amoco Cadiz* off Brittany, the *Eleni V* off Great Yarmouth and the *Christos Bitas* off the Welsh coast have severely tested those procedures. There are, however, other continuing causes for concern over possible emergencies which will come within the strategy of the national measures for biological conservation, including the hazards from accidental leakage of radio-nuclides into the environment.

British experience

The extent to which individual nations have shown concern for, and have reacted to, the measures outlined above has varied considerably. In Britain, a Wildlife Conservation Special Committee was set up by the Government to examine the needs for nature conservation. The report of this Committee, published in 1947, contained a list of proposed nature reserves where wildlife should be effectively studied and protected. The Committee also recommended the setting up of an official service to establish and maintain the reserves, to do research, and to advise generally on wildlife conservation. In 1949, the Nature Conservancy was created by Royal Charter as a new research council to undertake this work. In the same year, the National Parks Commission was set up to act as a corporate body to preserve and enhance the natural beauty of England and Wales, and similar proposals were made for Scotland.

Immediately after its creation, the Nature Conservancy began to survey and acquire many of the areas which had been proposed as nature reserves. As there were many other scientifically interesting places that could not be established formally as reserves, the Nature Conservancy assumed a statutory duty to notify sites of special scientific interest to planning authorities in order to give as much protection as possible to these other areas. This notification provided only a qualified protection, but ensured that, when proposals were made for developing a notified site, the planning authorities had to notify the Nature Conservancy so that its views could be taken into account. During its first ten years, the Conservancy had made good progress in establishing 84 National Nature Reserves, totalling some 56 000 hectares. It had also established an effective programme of research to study the changes that were taking place in the distribution and numbers of native wild plants and animals. Some of these changes were "natural" in the sense that they were the result of alterations over which man had little or no control, such as climate. Others were due to management or trends in land use which could only be altered with difficulty. Some of the factors were already understood, but many others still had to be discovered.

Until 1965, the Nature Conservancy operated as an independent research council specializing in terrestrial ecology but, in that year, the Natural Environment Research Council was set up by Royal Charter to co-ordinate the work of several separate bodies studying environmental matters, and the Nature Conservancy then became a component part of the new Council. In entering the broader sphere of conserving the resources of the natural environment for the wise use and long-term benefit of the community, the Conservancy moved beyond its original tasks of protecting wildlife and establishing nature reserves. By becoming a component part of the Natural Environment Research Council, it found itself more closely linked with other disciplines in the environmental sciences. This work was strengthened by the passing of the Countryside Act in 1968, and by the setting up of Countryside Commissions for England and Wales, and for Scotland. By 1970, there were 129 National Nature Reserves, with a total area of nearly 109 000 hectares. There were, in addition, over 2000 notified Sites of Special Scientific Interest.

In 1973, a further change occurred with the establishment of the Nature Conservancy Council as an independent statutory body whose members would be appointed by the Secretary of State for the Environment. The research scientists of the former Nature Conservancy remained with the Natural Environment Research Council, but continued to undertake research under contract for the newly established Council. The total number of National Nature Reserves is now 153, with a total area of 120 465 hectares. A survey of sites of conservation interest led to the publication of the Nature Conservation Review in 1977, giving a selection of biological sites of special importance for nature conservation in Britain. This report provides a list of the high-quality sites which would together provide an adequate series of National Nature Reserves for Britain. These sites are considered to be of national or even international importance for nature conservation, and it is hoped that they will be fully safeguarded in the future, in addition to the existing Nature Reserves.

The strength of wildlife conservation in Britain has always depended on a partnership between official agencies, voluntary bodies and individuals. This relationship has practical expression in the ownership and management of Nature Reserves, as well as in other wider aspects of conservation. Thus, well over half the total area of National Nature Reserves is managed under Nature Reserve Agreements with the owners and tenants of the land concerned. There are, in addition, 44 Local Nature Reserves declared by Local Authorities, totalling about 6000 hectares, and voluntary bodies manage some 37 000 hectares, nearly a quarter of the total area of Nature Reserves in Britain. Similarly, much of the relevant legislation already in force in Britain has been promoted by voluntary bodies such as the Royal Society for the Protection of Birds, the Council for Nature, and the Botanical Society of the British Isles. A substantial area of land in England and Wales belonging to the National Trust is managed as Nature Reserves, and nature conservation similarly forms a welcome element in the management of the policies of the National Trust for Scotland, four of whose properties (about 4500 hectares) have been declared Nature Reserves. There is also a close working relationship with other voluntary bodies, including the County Naturalist and Conservation Trusts in England and Wales and the Scottish Wildlife Trust.

European experience
Similar developments have taken place in European countries. Although names such as "National Park" and "Nature Reserve" have their equivalents in most European languages, differences in the meanings given to these terms can sometimes lead to misinterpretations. Thus, in Switzerland, Czechoslovakia and Poland, National Parks are managed rather like the larger National Nature Reserves in Britain, the first priority being the preservation of wildlife, with controlled public access, providing there is no danger to the scientific interest. In France, Spain, Italy, Austria and Germany, National Parks are mainly natural landscapes in which villages are absent or few and agriculture insignificant, but forestry, tourism and, in some cases, hunting may be important. The Nature Reserve as a category of protected land, with a strict interpretation given to the name, and meaning primarily the conservation of the scientific interest, is found only in Britain, the Netherlands, Czechoslovakia and Poland.

There are no two countries in Europe with identical organizations, and the British system is quite distinct from any other. Broadly, there are four groups of European countries with similar historical development of conservation organization, objectives and priorities, and the legislative powers available for such work. In Britain, Sweden, the Netherlands and Switzerland, influential and long-established voluntary societies pioneered the way for selecting and establishing Nature Reserves. When the respective governments were eventually persuaded to give official recognition and support to such work, much of the specialized knowledge was already available. In all four, the conservation of wildlife is the first objective, and the provision of outdoor amenities for the general public is a subsidiary function, but of varying importance according to the country. In Germany, Austria and Denmark, conservation has developed along different lines, despite the existence of well-supported and long-established voluntary societies for the protection of nature. While the planning and educational aspects of nature conservation are well developed, and many birds, mammals and plants are protected by existing legislation, the scientific conservation of wildlife is not a major part of the Nature Reserve programme. This approach to conservation is not ecological in the same sense as in Britain, and little research on scientific management is done in the Reserves. In Czechoslovakia and Poland, Nature Reserves and other types of protected land are all owned and controlled by the State, so that voluntary work is confined to education and publicity.

Since the war, scientific activities have received increasing support from governments, leading to a rapid development of studies in wildlife conservation and ecological research. In France, Belgium, Spain, Italy, Portugal and Greece, State support for conservation is generally weak or absent, and no strong voluntary movement has developed nationally. Public interest is slight and is not stimulated to appreciate the long-term value of an ecological approach to land use planning and conservation. In consequence, legislation to protect birds, mammals and plants is ineffective, except in a few local areas where the provincial authorities have strengthened them. In some Mediterranean countries, particularly France, the hunting interests act as a powerful political lobby and form an effective barrier to improved legislation for creating Nature Reserves and protecting wildlife.

North America

Experience in other continents varies enormously. In North America there is strong voluntary support for biological conservation, particularly in the better-educated sectors of the community. Division of responsibility for planning and land-use management between Federal and State authorities, however, makes many of the measures outlined above difficult to apply in a consistent sense. The Environmental Protection Act of the United States, on the other hand, has done a great deal to protect the biological environment against the more direct pressures of industrial and urban development.

Developing countries

In the developing nations, biological conservation is frequently seen as of secondary importance to the development of an adequate standard of living and a healthy economy. Nevertheless, considerable efforts are made in some countries to conserve wild animals and plants, as in the National Parks of East Africa and the Tiger Reserves of India.

International measures

There are, of course, international agencies concerned with conservation of the biological environment. These include the International Union for the Conservation of Nature and Natural Resources (IUCN); the International Council of Scientific Unions (ICSU); the United Nations Environment Programme (UNEP); and the United Nations Educational, Scientific and Cultural Organization (UNESCO).

International Union for the Conservation of Nature and Natural Resources (IUCN)

IUCN, in co-operation with the World Wildlife Fund, is engaged in a wide variety of projects related to the classification of the natural regions of the world, the preparation of lists and directories of protected areas and of endangered species of plants and animals throughout the world, and action to provide protection for threatened species.

IUCN was one of the sponsors of the Second World Conference on National Parks, held in Wyoming, USA, in September 1972. This conference devoted considerable attention to problems which relate, directly or indirectly, to the scientific problems of National Park planning and management. More recently, IUCN has proposed the formation of an international organization to provide professional services in natural resource management. The requirements for such services are now being met by individual experts, often recruited under technical assistance programmes, by a small number of non-profit-making service organizations, such as universities and, most frequently, by commercial consulting firms. To facilitate the provision of high-quality professional advice on matters which fall within its area of concern, IUCN is examining the feasibility of establishing an organization to provide such services.

While the organization would aim at being financially self-sustaining, its primary objective would be to serve the public interest. Potential clients of the organization might include governments, international development aid organizations, and national and multi-national industrial corporations.

International Council of Scientific Unions (ICSU)

ICSU is a non-government group of scientific organizations whose membership includes representatives from 66 national academies of science, 18 international unions, and 12 other bodies called scientific associates. To cover multidisciplinary activities which include the interests of several unions, ICSU has established ten scientific committees, of which the Scientific Committee on Problems of the Environment (SCOPE), founded in 1969, is one. Currently, representatives of 32 member countries and 15 unions and scientific communities participate in the work of SCOPE, which directs particular attention to the needs of developing countries.

The mandate of SCOPE is to assemble, review, and assess the information available on man-made environmental changes and the effects of these changes on man; to assess and evaluate the methodologies of measurement of environmental parameters; to provide an intelligence service on current research; and, by the recruitment of the best available scientific information and constructive thinking, to establish itself as a corpus of informed advice for the benefit of centres of fundamental research and of organizations and agencies engaged in studies of the environment. SCOPE, therefore, represents one of the important agencies which is currently concerned with problems of the conservation of the biological environment.

International Biological Programme (IBP)

In the early 1960s, ICSU launched an International Biological Programme (IBP), concerned with the "biological basis of productivity and human welfare", to promote the world-wide study of:

(*a*) organic matter production on the land, in fresh waters, and in the seas, and the potentialities and use of new and existing resources;
(*b*) human adaptability to changing conditions.

The field of study included directly or indirectly in these aims was very wide, but the programme was limited to those basic biological studies, related to productivity and human welfare, which could only properly be made by international co-operation.

In terms of human welfare, the value of IBP lay largely in the promotion of basic knowledge relevant to the needs of man. The rapid rate of increase in the world's population and the wide extent of malnutrition called for greatly increased food production, coupled with the rational use and conservation of natural resources. It was recognized that these aims could only be achieved through scientific knowledge which, in many fields of biology and in many parts of the world, was (and is) wholly inadequate. The importance of international co-operation in practical agriculture and public health had received recognition in the United Nations Food and Agriculture Organization (FAO) and in the World Health Organization (WHO), but there remained a need for strengthening international co-operation through biological studies upon which these activities depend. The international co-operation which was encouraged by IBP was of particular value to developing countries, in that it strengthened their scientific activity and provided new knowledge in fields highly relevant to their needs. IBP provided a unique opportunity for obtaining data about a large variety of environments with a degree of comparability, achieved by the application of internationally agreed methods, never previously obtained. The co-operation between scientists in different disciplines and countries stimulated new and productive lines of research. The national and international synthesis of research helped biologists to establish contact with others working on similar problems, and also drew attention to major gaps in knowledge.

The research carried out within the main research programmes varied considerably, and included the trial and development of instrumentation to determine the best methods to be adopted for internationally comparable studies, regionally replicated trials in order to provide data for synoptic studies on a regional or global scale, and research on new topics and problems which had hitherto proved intractable, together with the development of new lines of research from current activities. Much of the research on productivity focused on the pathways of energy and nutrients through ecological systems which have been described in outline above. Indeed, it was through IBP that many of the concepts which are now regarded as basic in ecology were developed, and these concepts are now having an important influence upon the planning and execution of new research. The spirit of co-operation which was built up through IBP has never been lost, and has been greatly extended in international research projects all over the world. The methodological studies, in particular, have helped to provide a basis for continued co-operation and comparison across the whole world.

Man and the Biosphere (MAB)
Today, many of the initiatives which were started in IBP and in other international programmes have been taken up and continued in

UNESCO'S Man and the Biosphere programme (MAB). MAB is an interdisciplinary programme of research which emphasizes an ecological approach to the study of interrelationships between man and his environment. It is implemented in close co-operation with the other organizations of the United Nations concerned with the environment and the competent international non-governmental organizations. It focuses on the general study of the structure and functioning of the biosphere and its ecological divisions, on research into changes brought about by man in the biosphere and its resources, and on the overall effects of these changes upon the human species itself.

The general objective of the programme is to develop a basis within the natural and social sciences for the rational use and conservation of the resources of the biosphere and for the improvement of the global relationship between man and the environment. It seeks also to predict the consequences of today's actions on tomorrow's world, and thereby to increase man's ability to manage efficiently the natural resources of the biosphere. With this general objective in mind, the programme is intended, more specifically, to develop a limited number of projects:

1. To identify and assess the changes in the biosphere resulting from man's activities and the effects of these changes on man.
2. To study and compare the structure, functioning and dynamics of natural, modified and managed ecosystems.
3. To study and compare the dynamic interrelationships between "natural" ecosystems and social and economic processes, and especially the impact of changes in human populations, settlement patterns and technology.
4. To develop ways and means to measure quantitative and qualitative changes in the environment in order to establish criteria for the national management of natural resources.
5. To help bring about greater global coherence of environmental research, by:
   (a) establishing comparable methods for collecting environmental data;
   (b) promoting the exchange and transfer of knowledge.
6. To promote the development and application of simulation and other techniques for the prediction of future changes.
7. To promote environmental education in its broadest sense.

The criteria for the selection of projects for the MAB programme are severe. Each project must provide information, through research, essential to rational decision-making about the use of natural resources. The necessary research must be feasible and likely to produce results in the short and medium term, of sufficient precision for the use that is to be made of them. Significant progress must be enhanced by international co-operation, the use of comparable methods, and the exchange of information. Further, each project must have intrinsic merit as a piece of scientific research. Research within the programme is focused on particular areas, the size and nature of which will depend on the geographical region, the problem being studied, and the resources available to the project. Thus, for example, an analysis of the grazing behaviour of wild herbivores will generally involve an area for research much larger than that required for a comparison of the growth of adjacent fields of wheat and clover. Some national projects inevitably focus attention on the impact of one type of stress on a single system or small area. Other projects deal with a mosaic of systems, each with its own structural and management characteristics. Still other projects take a broad physiographic or cultural region, such as a river basin or island, as the area for study.

Concentration of research on such areas as river basins illustrates well the interactions and interrelations that occur between ecosystems. For example, the water which falls from high mountains and high forests drains through the lowland forest, grazing land, agricultural systems on alluvial soils, and eventually to the lakes and rivers. Human activity modifies the interrelations between these systems, and this modification is reflected in changing patterns of productivity and in the transport of dissolved and suspended particles in water. The integrated effects of these changes are felt in estuaries, deltas and adjacent coastal waters. Such research is of immense significance in avoiding the kinds of mistakes which occurred through the construction of the Aswan High Dam.

The MAB programme as a whole may be perceived as a two-dimensional matrix in which projects representing geographical situations form one dimension, and projects representing processes or impacts of particular significance form the other dimension (figure 6.6). As a convention, the rows of the matrix are formed by the various geographically based projects, and the columns formed by process or impact-orientated projects. Thus, the projects on tropical and subtropical forest, temperate and Mediterranean forest, grazing lands, etc., represent the geographically based projects, while those concerned with the conservation of natural areas and the genetic material they contain, pest management and fertilizer use, genetic and demographic changes, perception of environmental quality, and pollution, represent the processes and impacts of particular significance.

The Man and the Biosphere programme may be seen as the latest, and very ambitious, attempt by man to promote an international understanding of the kinds of problems that are posed by the conservation of the biological environment. As a programme, essentially of research, it attempts to link the work and understanding of biologists with those of economists and social scientists and, for the first time, looks at the role of man in both creating and destroying his environment. In a sense, the MAB programme succeeds IBP, although it was not dependent upon it and is not organized in the same way. Certainly, much of the information was

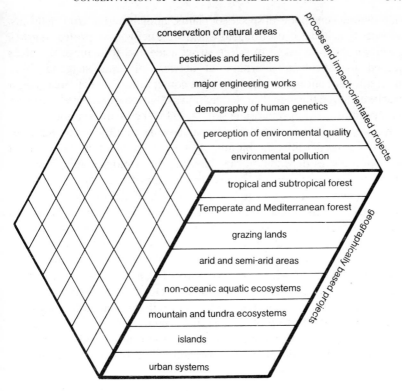

Figure 6.6 Matrix structure of MAB projects.

obtained through IBP has been invaluable in promoting the activities of MAB and has, indeed, been essential as a starting point for the research of MAB. Furthermore, many of the techniques that were developed through IBP have now been refined and developed as basic tools for MAB.

Biosphere reserves

The MAB programme has also recognized that the establishment of reserves, protected and managed in various ways, is of importance to mankind through the role these reserves can play in meeting scientific, economic, educational, cultural and recreational needs. Such areas are regarded as essential for research in ecosystems of various kinds and of fundamental importance for the MAB programme, since they represent base lines or standards against which change can be measured and the performance of other ecosystems judged. They also represent a means of maintaining gene pools of species of plants, animals and micro-organisms. In order to achieve a co-ordinated world-wide network of protected areas, governments were asked to select, to set aside and to manage the areas needed for such an international network. It was suggested that international concern for the long-term conservation of such areas might best be achieved through their designation as "biosphere reserves", together with promotion of approapriate standards for their conservation.

It is implicit in the concept of a global network that Biosphere Reserves should represent different biomes of the world, so that selection of appropriate sites depends on the classification of biomes and their sub-divisions. The selection of representative areas therefore involves two distinct stages. The first stage is the definition of the biomes, and their main subdivisions, which will be represented in the network. The second stage of the selection procedure is the examination of individual areas in order to decide which are appropriate examples of each desired ecological type. This selection involves the application of criteria some of which are essential, and others which are secondary. The essential criteria include representativeness, diversity, naturalness, and the effectiveness of the Reserve as a conservation unit. Secondary criteria include the degree of knowledge of the history of an area, the presence of rare or endangered species, and the relative value of an area for education and research. By 1977, a total of 130 Biosphere Reserves had been declared by 30 countries.

Genetics of biological conservation

From the outset, this review of the conservation of the biological environment has emphasized the complex interaction between man and his environment. Biological conservation involves more than the creation of visual and recreational amenity for the enjoyment of man—important though these qualities may be in other contexts. People, and particularly decision-making people, must soon come to understand that there are sound and compelling reasons for conserving natural areas and the plants and animals they contain. Recent work on the theory of ecosystem development, function and evolution, and studies of species diversity, provide a rationale for the "principle of co-existence between man and nature" which we have used as our definition of biological conservation.

Productivity
It has long been known that immature or early successional stages in the development of an ecosystem provide greater productivity for human

needs than late successional or mature stages, where harvestable productivity is reduced. Agriculture and forestry are based on this characteristic of ecosystems, although the origins of these practices are rooted in empirical discovery rather than in theoretical insight. Man must continue to maintain many ecosystems in immature stages in order to reap the benefits of high productivity. In contrast, the desirability of maintaining some ecosystems in a mature state does not seem to have been discovered empirically. Although their harvestable productivity is low, mature ecosystems serve a protective function. Characteristically, they have a high rate of exchange of carbon dioxide and oxygen, closed or nearly closed nutrient cycles, selection for quality rather than for quantity of organisms, and high efficiency in terms of the living matter supported by a given amount of energy.

The number of human beings in the biosphere has been greatly increased by the manipulation of ecosystems in the direction of increased productivity. Nevertheless, the mixing of managed land with land left in its natural state helps to increase the diversity of habitats which, in turn, often results in an increased variety of animals, and added benefits for agriculture. Agriculture and forestry, when well practised, are themselves dependent upon ecological principles, but some areas need to be retained as unmodified protected ecosystems for the healthy functioning of the biosphere as a whole.

Baselines

Protected areas may serve as a baseline against which change can be measured. To serve this function, of course, surveys must first be completed and a system of environmental monitoring established. Unmodified ecosystems which have not been significantly disturbed or altered by man, and particularly those in mature or climax stages, are most likely to fulfil this function as, when undisturbed, they change little with time and contain a rich assortment of organisms, some of which may serve as "early warning" indicators because of their susceptibility to particular pollutants or disturbances. For this reason, in many countries there is much interest in establishing automated meteorological stations on protected areas that have a high probability of permanence. Many of the older meteorological stations have been so modified by increasing human activity that records are no longer comparable from one decade to the next. There is also a regrettable tendency for man to concentrate his monitoring of pollution in urban fringe areas. In the measurement of sulphur pollution, for example, in Britain, very few of the monitoring stations have been established in rural environments.

Wildlife sanctuaries

Protected areas increasingly become the main sanctuaries for wild organisms. The reduction of species diversity deprives not only ourselves, but all future generations, of living resources essential for the economic, environmental, cultural and scientific existence of man. This attrition, proceeding with increasing rapidity, may become irreversible within a mere generation or two. Whatever our own attitudes may be, our responsibility to the future is to prevent the depletion or destruction of the genetic diversity of life, reduced though it may already be by the actions of our forefathers. There is a strong scientific justification for the preservation of rare and endangered life forms, because every species of living organism represents a store of genetic information which is irrevocably lost if it becomes extinct. We now know that such preservation is better accomplished in the context of the entire ecosystem to which a species belongs than in a zoo or a botanical garden.

What is at stake is not only the extinction of individual species, although the current rate of loss in many parts of the world is high (and accelerating), but the predictable destruction of habitats of what remains of the earth's natural and semi-natural communities and of the species that they include. Without deliberate protection, few of these communities have much chance of survival, and the shrinkage of undisturbed habitats does not offer much promise of evolutionary replacement. The pace at which these changes are taking place is unprecedented and, in the short space of two or three human generations, man's conflict with other organisms could virtually destroy a large proportion of the wild organisms that now remain. Man increasingly seeks to direct the evolution of organisms that are of use to him, so that the only organisms which will retain some evolutionary independence are those which he is unable to suppress.

Long-term values

Under the intense political and social pressures for an improved standard of living or, in the developed world, for the maintenance of the standard of living we have already achieved, people—and particularly those people determining policy—may object to the space and cost devoted to the conservation of the biological environment. In just the same way, those who travel in ships or aeroplanes may object to the cost of safety precautions, but these costs are still paid because it is known that disasters do occur from time to time. In the same way, we may need to use the genetic material maintained in the biological environment to offset disasters which have been brought about by man's misuse of his agricultural and forest ecosystems. This objection to the cost of conserving ecosystems becomes less of a problem when the values and aims of this conservation are more widely and effectively communicated to the general public. Moreover, in addition to the long-term values of the conservation of the biological environment, there are also immediate and short-term benefits. Many species of plants and animals that are now being used to improve cultivated or domesticated species, or to strengthen them against new outbreaks of insects or disease, are drawn from those natural and semi-natural environments which have remained relatively unmodified by man. Thus, Ethiopian strains of sorghum, rich in an essential protein-making chemical that other sorghums lack, have been used in improving sorghum as a food. The continuing existence of these strains has thus been of great importance in meeting the demand for more nutritious foods, a demand that will become even more important as the world population grows.

Genetic diversity

Recent scientific advances suggest that we may soon be able to synthesize genes, and the idea of creating a whole new organism is at least plausible. In terms of cost, measured either as money or as the input of energy, these alternatives are enormously more expensive than the conservation of genes and species that already exist. Newly synthesized genes and species may add biological understanding, but are unlikely to contribute greatly to the earth's biological resources. The new genes will not necessarily be adapted to any particular environment, and even a new species, while perhaps able to survive in the protected environment of the laboratory, may find much difficulty in fitting into communities of plants, animals and microorganisms which have evolved with each other, and in the natural environments, for many generations. Newly-created genes and species are most likely to die as a result of competition with native organisms, although the possibility that a new species will encounter no effective natural controls, and therefore prove destructive to many naturally evolved species, including possibly man, cannot be discounted.

An alternative approach to the conservation of biological organisms might be simply to store these organisms until they are again needed for the development of man's technology. Techniques for storage of organisms are already available, either for spores or seeds, or for whole organisms stored at very low temperatures. The development of whole organisms from single cells maintained in cell or tissue culture has already been accomplished for species of plants and may, in the future, be possible for all species, including animals. For such organisms, however, evolution ceases at the point at which they were stored. Species that fail to evolve in this way may well find themselves unable to compete when placed in an environment which is significantly different from that from which they were taken at the time of storage.

Paradoxically, the maintenance of the otherwise extinct species in zoos or botanical gardens poses the converse problem. After many generations, these token remnants of the original species find themselves adapted to, or even dependent upon, the environment of the zoological or the botanical garden in which they have been stored. The remaining, and apparently best, alternative appears to be the conservation of sufficient natural ecosystems to enable a significant proportion of the earth's plant and animal species to continue to exist and to evolve, surrounded by the environments of man's continually changing civilization.

Introductions

In many parts of the world, wild plant and animal species are of economic importance in pasture or forest production, for land reclamation, and as sources of various useful products. In many instances, important gains in production are obtained by making introductions of such organisms from other countries, so that a range of types adapted to a wide spectrum of ecological conditions is of the greatest importance for securing high levels of productivity. Major industries, such as the pastoral industries of Australia and New Zealand, and forest and fish production in many parts of the world, are largely or wholly based on such introductions from other parts of the world. A reservoir of ecological and genetic diversity in such species is therefore of the greatest importance for further development, and especially for the establishment of new industries in developing countries.

The focus of interest in the establishment of species in a new country is on the population, the ecotype, biotype, or provenance. The target of conservation is to secure the continuing existence of a broad range of genetically diverse populations, reflecting the diversity of conditions in which the species occurs. Populations in areas of natural distribution may be expected to have a wide spectrum of genetic diversity, but agriculturalists and foresters have come to recognize the value of genetic adaptations which are acquired in the countries of adoption. Thus, Mediterranean pasture plants were re-introduced to the countries of their origin after half a century of natural selection in Australia. Similarly, foresters frequently use locally adapted provenances of introduced species as a valuable source of future seed. In Britain, for example, where the forest industry is largely dependent on introduced species, seed orchards have been established from selected trees of the first-generation introductions.

Non-domestic animals

Many kinds of wild animals contribute to man's food supply, including most notably fishes, but also wild birds, mammals, reptiles and amphibia, as well as many invertebrates. Other wild animals produce furs, hides, plumes, etc., of direct economic value. In some countries, wildlife is a major source of animal protein in the diet of the people. One of the principal problems in maintaining populations of these economically important species, indeed, lies in control of exploitation, and in the special attention which needs to be given to the education of the public to the need for managing these species so as to ensure a continuing yield of their products. The fact that the animals are wild often suggests that they are common property, and therefore free to be captured and killed by anyone.

Generally speaking, the wild relatives of domestic animals are in a critical condition, and the survival of many is seriously endangered. For example, the wild relatives of the horse, ass, cow, sheep, buffalo, goat, yak, are all either endangered species or have endangered races. In many instances, populations have been further reduced by the destruction of their habitat, as well as by excessive exploitation. Interbreeding with domestic livestock, including feral populations of domestic breeds, is a threat to the genetic integrity of many wild populations. For example, the red deer in Britain have been markedly changed in genetic composition by the introduction of animals from elsewhere in the world, and by inter-breeding with animals which have escaped from wildlife parks and deer enclosures. While conservation of most wild populations of animals is best carried out by protecting the species within its native habitat, through establishment and strict protection of reserves, many of these species are now confined to small remnant areas within their original range, and there is the further need to establish new breeding populations through capture and transportation of the wild stock to secure habitats which are eco-logically similar to those already occupied. In certain cases, zoos can provide a temporary storage location for the restocking of natural areas.

Non-domestic plants
The wild relatives and sometimes the presumed origins of domestic plants are increasingly used in the improvement of cultivated stocks. These wild relatives provide important historical links with the early origins of crops and are much used in the study of the genetic and evolutionary relation-ships among the relatives of crop plants which are of actual or potential use in further crop improvement. Here, the main focus of concern is on the individual within a population, so that variation within and between populations is of direct relevance. Generally speaking, species of plants used in agriculture are relatively secure, although over-grazing and more intensive methods of cropping have endangered some Mediterranean species. Similarly, many important forest species are seriously endangered,

as are some groups of wild plants which are the relatives of several Mediterranean and near-Eastern vegetable crops, and various wild fruits in Malaysia.

Gene pools

There is no clear dividing line between wild species of plants and animals of direct economic importance and species of no apparent or immediate economic usefulness, since natural gene pools are likely to become increasingly important in human activities. Under the impact of population growth, technological innovation, and cultural and educational progress, landscape redevelopment for recreational and other purposes may assume considerable importance, even in the natural environments. Any such developments would require gene pools which far exceed the genetic resources of commercial suppliers of economically important plants and animals, or those of botanical and zoological gardens, and the gene pools of natural communities would then become important resources. With some exceptions, species which fall into this category must necessarily be conserved as parts of conserved ecosystems. Such an ecosystem must therefore contain either a representative or a unique biological community made up of a diversity of interacting wild species, and it must be of a size and shape to contain enough individuals of sufficient genetic diversity to ensure the long-term survival and evolution of one or more populations of the species in the ecosystem. Such conservation is only practicable and effective within communities adapted to their environment and to environmental change in a state of continuing evolution.

Man's role in biological conservation

Man has a critical role to play in biological conservation. He is part of the world ecosystem, but manipulates that ecosystem in a way in which no other animal can. Furthermore, he is now capable of harnessing enormous forces of energy in the modification or conversion of ecological systems into something new. Some of these modifications are not always what he expects, because the forces of ecological processes are equally great, sometimes operate in unexpected ways, and are unavoidable. As a result, these forces sometimes pass a harsh judgment on man, by the creation of deserts, by outbreaks of disease, by the existence of famine in a world of increasing agricultural production, by soil erosion and the loss of soil fertility, etc.

Land use

In urban environments, new cities and towns are created from brick, concrete and steel. In the perspective of historical time, these urban environments are ephemeral, sometimes to be abandoned and reclaimed. In these long processes of creation, abandonment and reclamation, there is a continual loss of valuable land which is brought into urban development, as man seeks new sites for his housing and industry, and extends the suburbs of existing towns and cities in peripheral and ribbon development. All these activities have their impact upon the rural environment, and lead to a steady loss of the valuable resource of land capable of growing crops, feeding food animals, or growing timber. Further areas of land resource are lost to mineral and industrial exploitation, and to the development of systems of communications by roads, railways, and the development of power lines and pipelines.

In the rural environments, agriculture is increasingly becoming more intensive, seeking higher yields through the use of specialized methods of production, in order to capitalize on short-term economic gains, despite the long-term inefficiencies of many of these developments. The requirements and taste for food become more sophisticated and more demanding. The more developed nations of the world, for example, eat large amounts of animal protein, despite the fact that even the most-advanced agricultural systems of meat rearing are inefficient in the conversion of energy and carbohydrates to proteins. Relatively short-term economic and social pressures lead to agricultural land being abandoned and reclaimed in ways which are obviously inefficient in the long term.

The exploitation of the world's forests, and the loss of forest land to cities, agriculture and, sometimes, to thoughtless destruction, makes feasible a world shortage of timber and wood fibre. In an attempt to make up for the exploitation of virgin forests, forestry has also suffered from intensification, leading to monocultures of exotic species which, while providing higher yields than uneven-aged mixed natural woodland, increase the danger of wholesale loss through pests and pathogens. The short rotations of modern commercial forestry capitalize on the higher productivity of immature ecosystems in ways that are inefficient in the use of the site resources and, if repeated, may deplete the stock of nutrients in the soil. The coppice and mixed-forest system of former times, while less productive under modern economic assessments, may, in the long run, prove more closely adapted to the stability and resilience of ecological systems.

The industrial and domestic demands for water in both the developed and the developing world show rapid increases, and these increases are frequently exaggerated by the wastefulness with which water is used. Part of the problem lies in the insistence, everywhere, of a supply of potable water for all purposes. Part of the problem lies in our unwillingness to measure and pay for the water that we use, a supply of drinkable water being regarded as a human right. In many parts of the world there is a distinct seasonality in the availability of water, leading to shortages at one time of the year, followed by flooding and flood damage in other seasons. Water is also linked with power generation, with agriculture, and with forestry in ways which are complex and not readily amenable to analysis.

However, perhaps one of the most intractable problems in the use of the rural environment is the tendency for the zoning of that environment to particular uses rather than the integration of those uses within a comprehensive land-use policy. Thus, in many parts of the world, land is devoted either to agriculture or to forestry, or to the exploitation of water resources, rather than these land uses being integrated within one particular area. In some countries, for example in Canada, land inventories have been made in order to identify the potential of land for particular uses, but such inventories have not necessarily led to an improved allocation of land to the best-integrated and optimum use. Problems of ownership and land tenure frequently make changes in land use almost impossible, and may place considerable constraints on the planning and development of national or regional land-use options.

Perception

Man's role in biological conservation is not, however, rooted solely in the way in which the real world changes in response to his activities. His role is also altered by his perception of the biological environment, and by the subjectively perceived world within which he lives. Descriptions and understandings of the biological environment are filtered by the senses, by minds, and by the means and media of expression, as well as by political and social institutions. Cultural patterns in different parts of the world also have their effect on the experience of environmental changes. The marked contrast between the pioneer and conservation ethics in North America illustrates clearly the change that has taken place on that continent. In the early days of the pioneers, land was considered to be expendable, and there was in any case so much of it that it hardly mattered what was done to any particular piece of land. The realization that the land resource is finite has now triggered off a keen awareness of the need for the conservation of the biological environment, and a demand for a more rational approach to the ways in which man's activities interact upon that environment. A recognition of the importance of conservation, initially of individual species and organisms, leads quickly to the concept of environmental quality which is now beginning to have a major impact upon those who make decisions about the environment, and on the citizens and voters who elect those decision-makers.

In part, the perception of the concept of environmental quality is related to the natural hazards of floods, earthquakes, tidal waves, and volcanic eruptions. Where such natural hazards occur regularly, they generate responses in the form of emergency arrangements and precautions in anticipation of the hazard. Nevertheless, these natural hazards can represent a major impediment to the economic and social advancement of developing countries, and can create conditions with which the political structure of a country is unable to deal. The result is a demand for external aid and for short-term measures which may further prejudice an effective long-term response to the existence of hazard.

Man-made hazards such as air and water pollution, and the introduction of toxic substances to food chains, take place in both developed and developing countries. The recognition of the hazard depends upon scientific information, and the use and understanding of such information. Frequently over-reaction to possible dangers leads to disillusionment with the scientific criteria, so that the hazard becomes discounted. An effective perception and response to the hazard therefore depends upon careful dissemination and communication of the basic scientific information, so as to obtain a measured and appropriate response.

Amenity and recreation
The increased demand for amenity and recreation which comes with improved standards of living has focused attention upon the rural environment as a series of landscapes. These landscapes have a direct relationship with past and present agriculture, and with other forms of land use, although neither visitors nor residents in the area may be aware of those relationships. Similarly, the interrelationships between buildings, spaces, open spaces and communications in urban areas, are important in the provision of amenity and recreation within such areas. Our improved perception of these relationships helps to form our ideas about degradation and improvement of both urban and rural areas, and creates an awareness of the importance of biological conservation which did not exist even a few years ago. The implementation of this awareness is a marked feature of our use and planning of land.

Environmental protection
One expression of the new awareness is the demand for studies of environmental impact of urban and industrial processes. The Environmental Protection Act of the United States, while not extensively copied by other countries, has generated a great deal of interest. There is now a fairly widespread appreciation that the efficient and economical attainment of environmental policy objectives requires that potential environmental

effects should be taken into consideration during the early stages of selecting and designing projects. For this reason, a comprehensive environmental policy needs to supplement controls over the environmental impacts of existing activities by an instrument ensuring that the environmental impacts of possible future activities are adequately taken into account in decision and planning processes. Such an instrument should satisfy two basic requirements:

1. It should provide for a systematic assessment of likely environmental impacts in a form appropriate to the activity to which it relates.
2. It should provide for the integration, at a sufficiently early stage, of the environmental impact assessment into the planning and decision process relating to that activity.

There are, however, some additional advantages:

1. Such provision may result in a more thorough and systematic appraisal of environmental impact than would otherwise tend to occur.
2. It may enable the general public to contribute, in a more informed and effective way, to the decision and planning processes for major developments having a potentially significant effect on their own future environment.
3. By introducing a significant element of external inspection, it may give greater assurance that environmental impacts will be assessed adequately and taken into account than could be achieved by a purely internal administrative commitment from each agency to take environmental effects into account in its decision-making.

Although many criticisms have been levelled at the particular system for environmental impact assessment which has been developed in the United States, it is clear that there is scope for a more regular and comprehensive assessment of the environmental impacts of new projects in other countries, and there is a widespread interest in the possibility of utilizing an environmental impact assessment system for this purpose. Environmental scientists will clearly have to be closely involved in the development of such systems. Land use planners, landscape architects and other environmental managers will no doubt play a central role in collecting, analysing and presenting much information but, increasingly, specialists in various aspects of ecology, and in pollution generation, dispersal and damage will be needed to ensure that an accurate and comprehensive picture is obtained. It is on this basic understanding of the underlying ecology of man's impact on his environment that effective assessment depends.

Systems ecology

In recent years there have been extensive studies of the behaviour of complex interacting systems in such fields as engineering, physiology, sociology, economics and geography. This work has depended heavily on the use of electronic computers, and is based on computer simulation or modelling. Some of the more ambitious studies have attempted to model complete regions, or even the whole world. Drawing on, and building upon, this diverse body of experience, progress has been made over the past ten years in a development of methods for understanding the dynamics of ecosystems and the impact of stresses upon them—including stresses generated by man. These methods are all broadly included within the term *systems ecology*.

Systems ecology is based on the assumption that the state of an ecosystem at any particular time can be expressed quantitatively, and that changes in the system can be described by mathematical expressions. Where this assumption applies, a quantitative knowledge of the state of the system at one time provides a basis for describing the state of the system at some later time. Commonly, the mathematical expressions describing the system are built into a computer program, and the computer is supplied with sufficient information to define the initial state of the system, the change with time of the expected impact upon the system, and the time interval over which a prediction is required. The computer then proceeds to a numerical solution of the mathematical expression, and reports the new values of the variables describing the current state of the system. The simulation model is a representation of the ecosystem and the processes within it simplified so that the behaviour of the system as a whole can be understood.

Once such a model has been built and tested, it can be used as a guide to the rational use and management of the ecological system it describes. If it is intended to modify the ecosystem in some way, the effects of this modification can be tested on the model far more quickly and cheaply than on the real-life system, and without the risk that unexpected effects may have embarrassing or irretrievable repercussions. If it is desired to manage the ecosystem to attain some specific goal, such as an increase in agricultural production without loss of conservation value, the model can be used to test all possible combinations of management practices, and so obtain the best results.

Such developments are as yet in their early stages. Ecosystem models have been built, and have been shown to behave realistically over periods of a few years. In a few cases, quantitative comparisons have been made between the results of computer models and observations in the real-life system they represent. We may predict with confidence that numerous such comparisons will be made during the coming years and that expertise in the field of systems ecology will rapidly expand. The need for assistance from this source in the rational use and conservation of natural resources will ensure that research in this field is directed towards practical problems rather than academic theories.

FURTHER READING

General discussions of ecology and biological conservation

Usher, M. B. (1973), *Biological Management and Conservation*, Chapman and Hall.
Edington, J. M. and M. A. (1977), *Ecology and Environmental Planning*, Chapman and Hall.

Classic review of conservation sites

Ratcliffe, D. A. (1977), *Nature Conservation Review: The Selection of Biological Sites of National Importance to Nature Conservation in Britain*, 2 volumes, Cambridge University Press.

Review of systems ecology

Frenkiel, F. N. and Goodall, D. W. (1978), *Simulation Modelling of Environmental Problems*, John Wiley.

Index